Science, Politics, Stem Cells and Genes
# California's War on Chronic Disease

**Other World Scientific Titles by the Author**

_Revolutionary Therapies: How the California Stem Cell Program Saved Lives, Eased Suffering — and Changed the Face of Medicine Forever_
ISBN: 978-981-121-328-1

_California Cures!: How the California Stem Cell Program is Fighting Your Incurable Disease!_
ISBN: 978-981-3231-36-8
ISBN: 978-981-3270-38-1 (pbk)

_Stem Cell Battles: Proposition 71 and Beyond: How Ordinary People Can Fight Back against the Crushing Burden of Chronic Disease — with a Posthumous Foreword by Christopher Reeve_
ISBN: 978-981-4644-01-3
ISBN: 978-981-4618-27-4 (pbk)

# Science, Politics, Stem Cells and Genes
# California's War on Chronic Disease

## Don C. Reed
Americans for Cures Foundation, USA

**World Scientific**

NEW JERSEY · LONDON · SINGAPORE · BEIJING · SHANGHAI · HONG KONG · TAIPEI · CHENNAI · TOKYO

*Published by*

World Scientific Publishing Co. Pte. Ltd.

5 Toh Tuck Link, Singapore 596224

*USA office:* 27 Warren Street, Suite 401-402, Hackensack, NJ 07601

*UK office:* 57 Shelton Street, Covent Garden, London WC2H 9HE

**British Library Cataloguing-in-Publication Data**
A catalogue record for this book is available from the British Library.

**SCIENCE, POLITICS, STEM CELLS AND GENES**
**California's War on Chronic Disease**

ISBN 978-981-126-141-1 (hardcover)
ISBN 978-981-126-214-2 (paperback)
ISBN 978-981-126-142-8 (ebook for institutions)
ISBN 978-981-126-143-5 (ebook for individuals)

For any available supplementary material, please visit
https://www.worldscientific.com/worldscibooks/10.1142/12997#t=suppl

Printed in Singapore

# Dedication

To the Caregivers:

You pay so much of the price of chronic disease and disability: from the endless physical backaches of lifting/transferring your loved one, to the loss of your personal life because of the caregiver's relentless schedule and chores; from the aching weight of inadequacy, that you cannot make your loved one well — to the worry of who will care for him or her when you are gone.

It is too much. Something must be done — which is why we fight for cures.

To all who are the givers of care, thank you. May you find relief in this lifetime.

# Foreword

Dear Friend and Fellow Patient Advocate:

If you have waded through my three other stem cell books, (STEM CELL BATTLES, CALIFORNIA CURES, and REVOLUTIONARY THERAPIES) feel free to skip the first several chapters of this one: they are background, synthesized from the first thousand pages of those three books.

If, however, you are among the 339 million Americans (and hoped-for new friends outside the United States) who have *not* read my books, the first few chapters are especially for you.

America is in a great war — not with guns or bombs, but a war nonetheless — to decide if science will be respected and involved. This is not small.

If the answer is "Yes!", we can handle the gigantic medical problems before us.

If "No!", I hate to imagine the world we will leave our children and grandchildren. The needless casualties will dwarf any previous plague. Even now, roughly half the population has at least one chronic disease or disability — and that is not even counting COVID--19.

We need science backed by serious money; we dare not cheap out on this one. Medically speaking, the troops need bullets.

California's stem cell program is the concrete realization of a great dream: that science will heal the sick and injured: that "incurable" diseases can be cured.

The alternative? We continue spending literally trillions of dollars on our sick every year, but not to cure them — just to maintain them, in expensive misery.

In 2004, California said "YES!" to three billion dollars in stem cell research. Now that funding is gone. Shall we build on its beginning, or say: "Enough! No more!"

Will we revive a spectacular program — for five and a half billion dollars?

Or shall it be allowed to quietly slip away?

Answering that question is the reason for this book.

# Contents

# 1 The Golden State, and Stem Cell Possibilities

California's Golden Gate Bridge: Linking an adventurous past and a shining future. (Wikipedia)

Long before I met Bob Klein, before I even knew what a stem cell was, I was a Private in the Army, my first day at Fort Monmouth, New Jersey, Winter, 1963.

It was bitter cold, seventeen degrees below zero, the radio said, and the Army in its wisdom sent us out to shovel snow. The wind blew so hard the glistening piles would be *whooshed* away, to instantly accumulate somewhere else. The moonlight made the snow look blue.

After several hours of this delight, an old sergeant suggested to the young lieutenant in charge that maybe we should go inside and warm up for a bit.

Back in the barracks, all of us shaking, shivering and cursing, somebody asked me, where was I from?

"California", I said — and everybody groaned.

I said, "What?" and my questioner said, "Are you one of those guys always says how wonderful California is?"

I thought about it for a minute and said:

"Well, it just is."

California is incredible. In physical beauty alone, it has no rival. Whatever scenery you want, it is in drivable distance in the Golden State. Want snow? Go north from the Bay Area, ski at Heavenly. Blazing desert? Turn left and head South, get all the sand and cactus you could possibly wish for in Death Valley.

It is also a place where *anything can happen*, if the citizenry says it should be so.

We have something called a Citizens' Initiative, or Proposition, whereby (with enough signatures and votes) you can literally change the state.

Propositions can be a mistake, as was (in my opinion) Proposition 13, which not only lowered taxes on the rich, but also made it extremely difficult to ever raise them again, requiring a 2/3 supermajority of the legislature.[1]

I remember my father coming home from work the day Proposition 13 passed.

"This is the end of California's leadership in education", said Dr. Charles H. Reed, Ph.D, Assistant Superintendent of Education, Mount Eden School District.

He was not wrong. By drastically cutting this source of revenue, Prop 13 devastated California education, dragging us down, from number one in the nation in terms of money spent per student — to forty-first.[2]

Citizens' initiatives can also be quite wonderful.

Like Proposition 71: the 2004 California Stem Cells for Research and Cures Act.

Should California spend $3,000,000,000.00 — three *billion* dollars (with a "B") — on stem cell research?

Or possibly — more?

---

[1] https://www.ppic.org/publication/proposition-13-40-years-later/
[2] https://www.sacbee.com/news/politics-government/capitol-alert/article226179020.html

# 2 The Dream I Wish Had Not Come True

On September 10th, 1994, Roman Reed became paralyzed in a college football accident. (San Jose Mercury-News)

I sat up in the moonlight, sobbing, shaking, tears running down my face.

Jeannie was awake instantly, patting my back, pulling me close, asking what happened, what was the matter? Did I have a bad dream?

Yes, just a dream.

But such a terrible one: our son, Roman, was lying motionless on green grass. He was wearing football gear, but his *helmeted head was separate from his body.* And the head spoke to me, saying:

"It's all right, Dad, I'm okay, it's all right", over and over.

I do not believe in predestination, that our lives are laid out before us. I believed then, and still do today, that we are the writers of our lives.

But because of that dream, and what came after, I am no longer quite so sure.

Christopher and Dana Reeve: their legacy lives on, in the CDRF Foundation (website), led now by Maggie Goldberg, CEO, veteran patient advocate.

In Roman'a darkest hour, Senator Dianne Feinstein (website) and Representative Pete Stark (Wikipedia) were there for us.

Two weeks later, September 10th, 1994, Roman Reed broke his neck in a football game, at Chabot College, Hayward, California.

I remember how carefully we picked him up, the line of coaches and fathers, and put him in the back of the Emergency vehicle.

At the hospital, things got worse.

"Your son will never walk again, never close his fingers, and almost certainly never father a child," said the doctor, with Roman in the hospital bed before him.

To illustrate his point, the doctor lifted up our son's hand, and told him to "keep it in the air". When he let go, Roman's hand fell so fast it slapped his own face.

"See?", the doctor said, "No triceps function".

Jeannie grabbed the doctor's shirtfront and might have hit him, had I not intervened. Then Roman asked the doctor to leave, in words not repeatable here.

Later, when I was sitting outside Roman's hospital room, trying to think of something to do, a book flew through the air like a Frisbee.

"Get that for my brother," snapped daughter Desiree. Book and problem landed in my lap.

One of Roman's coaches had given her the book, "RISE AND WALK: the Dennis Byrd Story". A football player on the New York Jets team had been paralyzed at the same injury level as Roman: crushing the fifth and sixth cervical vertebrae (C-5,6) while making a football tackle. But Dennis Byrd walked again; he had been given a drug called Sygen, made of dried cow brains. He was not healed, not completely, but he could walk: his situation was definitely improved.[1]

A small miracle: the night nurse told us there was a clinical trial going on right now, to fight paralysis, testing that same drug!

I called up the hospital in charge, asking if we could bring Roman, get him involved. But it was midnight, flooding down rain, and the nurse said no, you will just make him sick, bring him in the morning, that will be fine.

But it was not fine. When I called in the morning for instructions, they said we missed the cutoff time by one hour. The study required the drug to be administered within 72 hours after the injury, and this, apparently, was hour 73. By the passing of a little piece of time, we had lost our eligibility.

---

[1] RISE AND WALK: the Trial and Triumph of Dennis Byrd, by Mike D'Orso and Dennis Byrd, published by Harper/Collins, 1993

But still, might there be another way to get the medicine? I located the drug's inventor, Dr. Fred Geisler, at the University of Chicago. I asked him if the Sygen could still work, even past the 72 hour line. He said it might.

But we would have to get permission from the FDA first, also a prescription for it, after which we could order the drug from Switzerland.

Representative Fortney "Pete" Stark contacted the FDA on our behalf, obtaining a "compassionate use" permission.

But the doctors said no. I went from floor to floor of Roman's hospital, asking for help.

And then — Dr. Chi Chen Mao.

"I no think it can work," he said, in a heavy Mandarin accent, "But I father too, understand you must try." And he signed the prescription slip for an experimental medicine.

The cost was $35 per dose, from a company called Fidia. I ordered several boxes from Switzerland.

Roman endured 39 days of rehabilitation, all our insurance carrier would provide.

Safety-belted on a motorized bed, he was slowly rotated to vertical, inches at a time. It took almost a week to get him used to being fully upright. It was painful, he said, like an ice cream headache only worse, but it slowly restored the blood flow to near normal.

On almost our last day of rehab, the boxes of Sygen arrived. Roman's girlfriend Terri was shown how to give the injections.

We took Roman to another therapy center, in Southern California.

I have nothing good to say about that place, but it was all we could find. For months, Terri and I worked out with him at what we called "the gym", during the day, and in the hotel room at night.

How does a paralyzed person exercise? Teamwork. The able-bodied partner lifts a limb, doing as little as possible, while the paralyzed person helps as much he/she can. Although nothing much seems to be happening, it is exhausting for all.

Albert Matej, a massage expert, took us under his wing. He had worked a lot with paralyzed people, as well as movie stars like Bob Hope, and athletes like Junior Seau. Albert helped us build a "Stryker Bed", a bunk bed with loops of rubber cords hanging down. You put the person's leg in the loops, and it supported the weight, saving your back from so much lifting.

We gave Roman two injections a day, $35 a shot. The rehab center was $5,000 a month. We borrowed, going deep in debt.

Then came a tragedy — Christopher "Superman" Reeve's accident. Thrown from his horse, Reeve broke his neck at the top of his spine, cervical one (C-1), just below the skull. There is no place worse to injure the spine. For a long time, he could not even breathe on his own.

When we first heard the news, there was an ugly thudding noise — it was Roman, banging his wrists together, which is how quadriplegics applaud.

"Don't you know what that means to him?" I snapped at Roman.

"He is a superstar," said Roman, "Paralysis just got a face."

And so it proved, in a way none of us could have predicted.

Some in Reeve's family blamed the horse for stopping short and pitching the rider. They wanted to shoot the animal.

But Christopher said it was an accident, and would not let the horse be put down.

A woman named Joan Irvine Smith heard about Christopher's act of mercy. The Irvine family is very significant in Southern California, with a city named after them — and Joan Irvine Smith loved horses.

Reeve's act touched her so much, she gave him a million dollars to start a paralysis research lab.

Meanwhile, my wife changed her name. Roman told her he always liked her middle name, Gloria, and would she mind going by that? Right now she would do anything to make him smile. She did not realize he had done it to distract her from the worry; every friend she met, she had to explain why she was now Gloria, instead of Jeannie.

She also had our house extended, adding on several rooms (more debt) for adapted living. The entire back wall of our home was removed, and plastic sheeting tacked on. At night the winds whistled through splits in the hanging plastic.

Representative Pete Stark continued his kindness. He donated $500 for Roman's therapy (as did U.S. Senator Dianne Feinstein; and children's author Beverly Cleary sent us $100). Amazingly, Congressman Pete also got us two months free rent in a hotel attached to a rehab hospital. There Roman learned practical skills, like how to drive an adapted van.

With the help of Sygen and/or the exercise, his triceps function returned.

But when we got home, it was to a huge argument.

Roman would need an attendant for the rest of his life. That would be Terri, if she stayed, and became his wife, and also me. I figured that was a lot, and all that could reasonably be expected. Other than that,

I wanted to stop — no more impossible missions to find experimental medicine, no more complicated research papers, trying to decipher Latin medical terms.

"I got your triceps back!", I said to him, "You can drive an adapted van!"

"That's good," said Roman, "Keep going. Find a way. *Fix this!*"

Fix paralysis, incurable since time began?

The first historical mention of paralysis was on the wall of an Egyptian tomb.[2] In pictograms, it said: "Of paralyzed soldiers, deny them water, let them die — there is nothing that can be done."

From that day to this, no one has been cured of paralysis — and he wanted me to "fix it"?

"I want my life back!" I wailed.

"You don't have a life!" said Roman.

---

[2] "QUEST FOR CURE", by Sam Maddox, published by the Paralyzed Veterans Association. 1993

# 3 The Shot Policeman and the Walk-again Rat

Assemblyman John Dutra, author of AB 750, the Roman Reed Spinal Cord Injury Research Act of 1999. (John Dutra website)

One rainy New York night, patrolman Paul Richter asked a suspicious someone to open his car's trunk. The man said sure, no problem, turned the key in the lock, clicked it open — and whipped out a shotgun — blasting away at point blank range. Shot in the chest, Richter went down. But he drew his weapon and was firing as he fell.

The gun-runner got away temporarily, but was caught and convicted later on. In jail, he had a heart attack sitting on the toilet, and died.

As for Paul Richter, the shotgun pellets went through his torso. Some grazed his spine. He was not paralyzed, but had extensive nerve damage and continual pain, and was forced into early retirement.

Hold that name: Paul Richter.

Roman worked tirelessly to inspire research legislation. (reeve.uci.edu)

Spinal cord injury expert Os Steward provided leadership to the Roman Reed Lab. (provost. uci.edu)

An early endorser of "Roman's Law" was Bill "Superfoot" Wallace, world champion kickboxer, movie star, and trainer of Elvis Presley. (Wikipedia).

In my early years as a professional diver for Marine World, (1972–86), an aquarium-zoo in Redwood City, California I would sometimes have to turn over a living whale, whether orca (technically a giant dolphin) or pilot whale, a 20 foot long black tube of muscle. When this happened, for the animal's medical needs, I would study the situation first, try and figure out exactly where to push or pull. Once I had a point of leverage, then I could make a difference.

Trying to find leverage on paralysis, I attended a convention on spinal cord injury. It was a beautiful oceanic setting, Asilomar in Pacific Grove, California, but intended for scientists, and I could not understand what they were saying.

I kept nodding off from brain overload. The researchers' big words just would not stick in my small mind. I asked so many questions, the organizer told me no more, or I would have to leave.

Afterward I approached a scientist and said:

"I don't understand what you guys are talking about, just that it is important. So I can't help you with the science. But is there anything else you need?"

"Money," he said, without hesitation, "If I don't get grants, I can't put food on the table, and the research stops."

Money? How hard could that be?

It was like one of those old black and white movies where the kids need to raise money for a good cause, and Judy Garland says: "Let's put on a show!"

With the help of my 8th grade multicultural club, True Colors, I wrote, produced and put on several plays, including: "A NIGHT FOR NO MEXICAN TEARS", about the Mexican-American revolutionary, Juan Cortina. Our show featured a cap pistol battle with 40 toy guns (imagine the racket!), and a live horse ridden across the stage, when the hero returns at the end. We also sold snacks.

In the intermission, the PTA parents came on stage and counted the money. We raised $4,000, and sent it to Christopher Reeve.

He sent back a wonderful dictated letter, one sentence of which stayed with me ever since. He said:

"One day, Roman and I will stand up from our wheelchairs, and walk away from them forever."

Meanwhile, Patrolman Paul Richter was still fighting: only now for paralysis cure.

He thought: the number one cause of Spinal Cord Injury (SCI) is car crash. Why not tack a fine onto traffic tickets, raise money for research for cure?

"Uncle Paul", as he is known by his friends, meaning anybody who knows him for five minutes, approached the Governor of New York, George Pataki, and asked if he would support a traffic ticket add-on.

Sure, said Pataki — if Richter got approval from 100% of the legislature, both sides of the aisle, every Republican and every Democrat.

Which was of course impossible.

Except he did it, $15 add-on per ticket! Could California could do the same?

Using my typewriter (I had no computer back then), I wrote every member of the California Assembly and Senate, asking them to support a program to fight paralysis. (Gloria was not happy about all those stamps!) I got back about six responses. But one of them gave me a scribble of advice: go local. So I did, sending a more individualized letter to our district's Assemblyperson.

Silence… and then one day, in the middle of my 8th grade English class —

"This is John Dutra," said the voice on the phone, "Your assemblyman."

He had a kind voice. He apologized for disturbing me in class time, he said, but in his office they had read my letter out loud, and some of the staff were crying.

Would I mind, he said, if they named a law after my son? It would be Assembly Bill 750: the Roman Reed Spinal Cord Injury Research Act of 1999.

Would I mind?

As I hung up the phone, I screamed, and hit the wall hard with the base of my fist, and ran out of my classroom, down to the office.

I got on the intercom and told the whole school what had just happened. They had no idea what I was talking about, but they gathered I was happy, so that was okay.

For that matter, *I* had no idea what I was talking about! The only thing I knew about bills and laws was that song, "I'm just a bill!" from School House Rock!

But Dutra knew. He had us get together with Ryan Spencer, his Legislative Aide, who would tell us when the committee meetings were — at which we would speak. Ryan Spencer was a champion powerlifter, axe-handle wide across the shoulders.

Legislative aides like Ryan are the folks who get things done. Quietly, behind the scenes, an aide can make or break your bill, by ignoring or encouraging it.

We had to contact the aides of each committee member. They would know if the bill was something their leader could support. We learned to walk into an Assembly person's office, and just ask: who is the aide in charge of health bills?

Roman would drive his van from Fremont to Sacramento on a moment's notice. He loved to speak at hearings and would always ask me to write him something — which he would cheerfully ignore. But he had the unlearnable gift of charisma. People wanted to hear what he had to say, so it worked out.

We made friends like Karen Miner, a fragile wisp of a woman, whose spirit was not bound by her wheelchair.

Example: later on, she would become webmistress for my column, STEM CELL BATTLES. When she fell off her wheelchair, and Karen broke her collarbone, I thought that was the end of her involvement, but she said, no, she would just use her other arm.

She introduced me to Susan Rotchy and Fran Lopes, more wheelchair warriors, and they knew Marco and Shelly Sorani, and so the bill began to grow.

We picked up surprising endorsements: like Bill "Superfoot" Wallace, world champion karate star. You might have seen him in movies where he fought Jackie Chan, or Chuck Norris. They called him Superfoot because of his phenomenal kicking speed and balance. He had badly broken one knee, and could not kick with it. So he used the "bad" leg exclusively for balance, and endlessly kicked with the other — and became World Champion.

Iraq warrior U. S. General "Stormin' Norman" Schwartzkopf gave us his personal endorsement, saying:

"I have led troops into many battles, but never one so important as the fight against paralysis."

We also picked up opposition. The American Automobile Association (AAA) took an "oppose" position. They did not want their drivers inconvenienced, apparently not even the bad ones.

Karen and I organized a letter-writing campaign. So many letters were pushed through the slot in the lobbyist's office door, it was hard for her to open it! The AAA changed its position from "oppose" to "neutral".

We were also opposed by the U.S. Council of Catholic Bishops — because we supported embryonic stem cell research. There was nothing we could do about that, except to just go ahead, so we did.

In an amazing committee meeting, Senator John Burton recommended our bill for $19 million a year from the General Fund — *the Republicans agreed* — we had it made, we thought.

And then came the energy crisis.

People's cars lined up for blocks, just to get enough gas to go to work.

In Sacramento, every bill which had a cost was put into the "suspense file" — to be reconsidered — and there our bill died.

Naturally, we tried again, this time asking for just one million dollars a year.

We won! We went all the way through both houses of the California legislature, the Assembly and Senate. Governor Gray Davis signed it. We had a law!

March 1, 2002. I held in my hand a rat which had been paralyzed, but which now walked again. I could feel the tiny muscles struggling to be free. When I set her down, she scampered, tail-high, across the plastic swimming pool play area.

The rat's name was Fighter, and she had been given embryonic stem cells, an experiment performed by Dr. Hans Keirstead, the first of its kind in the world. I had seen video of her when she was paralyzed, dragging her useless hindquarters like luggage. It was hard to watch. But now, she ran...

This happened at Opening Day for the Roman Reed Laboratory, headquartered in the Reeve-Irvine Research Center. In charge was Dr. Oswald Steward, a cheerful scientist with many years of experience.

Christopher "Superman" Reeve spoke to us on the speaker phone, saying: "Oh, to be a rat this day!"

Gloria got so excited, she photographed the phone!

It was wonderful — but a million dollars a year? That was roughly the cost faced by one newly-injured patient, in his or her first year...

We could provide only the smallest of grants. These were useful — we helped pay for Keirstead's early work, for example — but couch change compared to the need.

The only way Roman could be healed, would be if the whole field of regenerative medicine moved forward.

That was the place to push — but how?

# 4 The Importance of Pizza (Boxes)

Bob Klein exudes energy, like he could light up a room. (alchetron.com)

"He is a smiling pugilist" –Marianne Moore, of Muhammad Ali

"Bob will see you now," said the secretary.

Behind a big table was a broad-shouldered man with a craggy face. He held up one finger, continued scribbling something illegible with a blue felt-point pen. His handwriting is as bad as mine, I thought. Well, maybe not that bad. I had proposed to Jeannie in a hand-written letter, which she could not read, "Looks like chicken-scratch to me," her mother said, and they threw my proposal away...

It occurred to me that Bob's last name, Klein, was the German word for little, which fit; he was fighting diseases with the tiniest of weapons: stem cell research.

He was attempting to raise money for medical research — as I had done — but his technique was different, and his "ask" was for billions.

Raising money for a good cause was nothing new for Robert N. Klein II. He had helped write and establish the California Housing Authority, which provided low-cost loans to middle and lower income citizens.

He helped JDRF International (formerly the Juvenile Diabetes Research Foundation) raise $1.5 billion from the US government — just before the invasion of Iraq, when all available funding was tied up in the preparations for war. President George W. Bush refused to sign any bill unless it had 100% support from both sides of the aisle, plainly impossible. But (like Paul Richter's long shot) Bob and JDRF's Larry Soler made it happen.

He had been approached by a patient advocate, who asked, would he commit to raise a billion dollars — for stem cell research?

"No", said Bob.

"Too much?" asked the advocate.

"Not enough," said Bob Klein, "*It has to be at least three billion dollars.*"

And now Bob Klein put down his pen, leaned back in his chair, hands behind his head. He gave a great beaming smile: as if everything was cleared off his calendar, there was nothing in the world he wanted more than to meet with me, and he had endless time for it. That was an illusion, I knew. Here was one of the busiest men in the world, and I had better not waste time stuttering.

"Tell me", he said, so I did, the whole Roman Reed funding story.

I had also provided patient advocate backup for California Senator Deborah V. Ortiz, as she passed the world's first stem cell legislation, three laws to make the research legal. None of them came with money, but still they allowed the research to go forward: permission bills.

Then I stopped. It was all I had to say. Was it enough? My life was hanging in the balance. Here was the man who could change the world. I *had* to work with him.

"Let's go meet the others," said Bob Klein.

Autographed paintings from Disney's TOY STORY smiled down on us from the walls. The smell of pizza called. (In high school I was three-time pizza eating champion, once devouring an entire combination pizza, large, in less than three minutes.) From a distance came the sounds of cheerful chaos.

The door opened into what looked like a college lunchroom, jam-packed with busy cheerful people, everybody talking at once. On the sinks were stacks of empty pizza boxes, as well as some with contents still intact.

Behind one desk was a giant cardboard thermometer, with numbers on it — 5, 10, 15, 20, 25, 30 at the top.

Seated there was a young woman, who smiled and waved at Bob. She had a phone snugged to her ear.

"Meet Amy Lewis," said Bob, "She's in charge of fund-raising for the campaign."

People saw Bob. Heads turned, as when a lion approaches the waterhole. They were plainly remembering things they needed to ask Bob. A queue began to form.

"Bob, your ten o'clock!" said a tall woman with green eyes, and took Bob out of the room. He did not look back. He had moved on.

"The thermometer shows the campaign budget," said another tall woman, "I'm Amy Daly. That was Amy DuRoss who just left, and you already met Amy Lewis." The woman beneath the money thermometer smiled and waved again. "Since there are three Amies, we go by last names — call me Daly."

"How much do you have so far?" The thermometer was blank, un-colored, except for a little green smudge at the bottom.

"Just 100,000", said Daly, "Bob's gift to get us started."

I had never been on a campaign with an actual budget. We always operated on the zero-budget-option, meaning we could do anything, as long as it cost nothing, or we paid for it ourselves.

"What happens next?"

"Signatures," said Daly, "We need at least 565,000, plus maybe 100,000 more as a buffer in case of invalid ballots," She pointed at some stacks of petitions.

"Take some with you when you go," she said.

"But is there something I can do — right now?"

She looked toward the pizza boxes on the sink.

"We ordered clipboards for the signature gatherers, but they have not arrived. If somebody was to take a razor blade and cut up those boxes, that would give us cardboard backing for the signature gatherers...."

I went home with a blood blister on my thumb. The boxes were gone. So was the pizza.

# 5 Please Help Change the World!

Bob and his inspirations: Jordan, Robert and Lauren.

I did not think Gloria Jean could help very much on signature gathering. After all, I had been studying up on stem cell research, especially the "Message Points" sheet, and she had not.

Still I was glad for her company, not to mention I tried not to argue with her needlessly — my wife is a little bit ferocious.

We had our signature-gathering kit: a folding card table (one leg was bent, but usable), two folding chairs, blank petitions, sheets of message points, a dozen ballpoint pens, and a sign which read "STEM CELLS ON THE BALLOT?"

It did not begin well. We set up in front of a grocery store, and were immediately approached by a passer-by. She read the sign, looked at us suspiciously, asked what we were doing, and then ran into to the store, emerging with a manager, who evicted us.

"Tattle-tale!", snapped Gloria Jean, "I hope _you_ never have a child who is paralyzed!"

The second try we asked the manager first. He said "No, no, too controversial!"

But the third manager had a sister with diabetes. He made us welcome, helped us set up, even gave us some bottled water!

Seating myself comfortably, I glanced over my message points, waited for someone to read the sign and come over. A stream of people passed. Some read the sign — I could see their lips moving — but they did not stop.

"This isn't working, " said Gloria Jean. And she stood up.

She did not exactly tackle her first prospective customer, but she did take his sleeve, and his face changed expression as she maneuvered him to the table.

"This gentleman wants to get stem cells on the ballot", she said.

I went into my memorized lecture.

"Stem cells are unspecialized cells that can generate healthy new tissues and organs," I said, "Possible treatments for many diseases including — "

Gloria had now sent two more people over, making a line. The first person shifted foot to foot.

"Cancer, heart disease, diabetes, Alzheimer's, Parkinson's, HIV/AIDS, multiple sclerosis, Lou Gehrig's disease, spinal cord injury — "

"He just wants to sign the petition, Hon!", called Gloria Jean.

"Yes, dear," I sighed, handing the man a pen.

We later watched a professional signature-gatherer at work. It was a revelation. He had "the ask" down to four words. "Support stem cell research?" he would say. If they said yes, he gave them the clipboard and pen. If they said no, he moved on.

In terms of signature gathering, there seemed to be four categories.

Category one: YES — already convinced, let him or her sign.

Category two: MAYBE — give a short talk, 2-3 sentences, no more — no sense lecturing a "maybe", while two or three "yes" folks walked by.

Category three: NO — if someone was plainly against the research, don't waste your time, or theirs.

There was also a fourth category.

I was tabling alone at one of my favorite places on earth, in front of Sather Gate, doorway to UC Berkeley, the intersection of University and Telegraph. I was born and raised in Berkeley, and love it there.

But as I answered questions from one "maybe" person, I noticed an elderly couple reading my sign and apparently discussing it. Then the woman's lips compressed to a thin tight line, and they stopped talking: maybe irritated having to wait, I thought. Hurrying my current customer to finish the form, I rushed to the waiting couple. Did they want to help get stem cells on the ballot?

The woman *spat* on me. I heard the sound, saw the white twist fly through the air, landing on my shoe.

She looked at me. I looked at her. I saw her face moving, like in a silent movie, and she was saying something, but I did not really hear. Only the anger registered.

Afterwards I thought of many cutting things I could have said, should have said, but didn't. The husband steered her away. She seemed content, having delivered her message.

I had wondered, what would it be like to confront the naked face of hate. Would it be terrifying, a person utterly devoid of reason? Or infuriating, someone who would deny research which might help my paralyzed son?

What I felt, (but did not say), was: Lady, you are a little bit crazy.

The spit dried on my shoe, and I went on gathering signatures.

I was glad she was not carrying a gun.

A couple weeks later, when I showed up at the campaign office, a security guard was blocking the door. What was this about, I asked, as the guard located my name on a list.

"Didn't you hear?", he said, "They blew up a stem cell lab."

# 6 "But Where Do You Keep the Babies?"

"Let's collect a million signatures!" said Bob Klein. (Chronicle/Shelly Eades)

"Police Confirm Pipe Bomb Blast at Stem Cell Lab", the headline read.

"An explosion that blew out a number of windows at a Boston-area laboratory specializing in stem cell research was caused by a pipe bomb, local police said...No one was wounded in the blast at Watertown, Massachusetts-based (company name withheld), which is working on cures for diabetes and liver disorders."[1]

When questioned, opponents of research disclaimed all responsibility. Nothing to do with them, they said, just some disgruntled former employee.

---

[1] "Police Confirm Pipe Bomb Blast at Stem Cell Lab", Boston, Reuters, Yahoo News, August 28, 2004

I was not so sure. Where did people like the shoe-spitting lady get their ideas? Seemed to me, they were being pumped full of poisonous misinformation.

Imagine someone emotionally on the edge, who goes around brooding, mumbling to themselves. How might he or she feel (and perhaps act) if someone told them stem cell research is...*murder*?

Misunderstandings are possible under the best of circumstances.

One friend of the family asked me:

"Where do you keep the babies?"

"What babies?"

"You know, the ones you get spare parts from, like they told us about in church?"

She had been hearing statements like the following:

"Embryonic stem cell research is...indisputably killing...Living human beings should not be used for harmful research without their consent."[2]

To understand why the above is false, we need to see how embryonic stem cells are actually made.

First, imagine a couple unable to produce a child the usual way. They might turn to the In Vitro Fertilization (IVF) method, which has been successful for more than five million babies born, all around the world.[3]

The husband has the easy work. He just donates sperm. That's it. The wife's part is far more serious. She takes hormone injections to increase the number of eggs within her body, and then has them removed: an oocyte-retrieval operation.

The sperm and eggs are put in a dish of saltwater. Her blastocysts (microscopically tiny fertilized eggs) come together, generally 15-20. Of these, the healthiest 1 or 2 are placed inside the woman's uterus. Hopefully, one will implant itself in the wall of the womb and become a baby. We wish them all good fortune.

But what about the other fertilized eggs, the less viable ones, blastocysts that are not used? Here is my non-scientist's explanation.

Some may be frozen and stored (at an ongoing expense, $500 — $1,000 a year ), or donated to other couples, or sold through a donation agency.

---

[2] https://www.csmonitor.com/2001/0313/p2s1.html
[3] https://www.sciencedaily.com/releases/2018/07/180703084127.htm

Most are discarded: flushed away, or added to the landfill. Others (perhaps 20%) are "abandoned" and simply remain in a frozen state for years.[4]

And some — (after the decision is made to discard) may be donated to research. These bits of tissue, are taken apart under a microscope (there can be no pain — no nerves) and the stem cells are removed. These look like microscopic basketballs. Now there is a stem cell "line", which may help in the struggle to find cures: to end suffering and save lives.

That's it, the whole process — so, where is the "living human being"?

You may look as long as you wish, and you will never find one. There is no implantation: no womb.

It is biologically impossible to make a baby without implantation in the womb.

No implantation, no baby.

(Birth control pills work by preventing implantation.)

Stem cell research, as the name implies, is cells, cells, nothing but cells: bits of tissue that would otherwise have been thrown away.

How were we patient advocates to deal with outright attacks or misinformation?

"Provide accurate information," we were told, "and if that does not work, walk away." Nothing is gained by debating someone whose mind is already made up.

Once, Bob was on the radio, and a caller insulted him: "You are like Joseph Mengele, the Nazi Angel of Death!"

I wanted to punch the radio. Mengele was one of the most evil men in the history of the world; he would inject blue dye into the eyeballs of concentration camp inmates, without anesthesia.

But Bob just said, "Thank you for raising that important point," as if the man had done him a favor — and went on to say what he had come to say.

At the end, the caller said he had never thought of it like that before!

We cannot persuade everyone. The folks with their minds made up? If they are on our side, no need to convince them, other than to encourage

---

[4] https://www.nbcnews.com/health/features/nation-s-fertility-clinics-struggle-growing-number-abandoned-embryos-n1040806

them to keep up the good work. Those who strongly opposed us — let them be. Do not engage with them, unless in an actual public debate.

But those in the middle, not quite convinced, one way or another? These were the people we had to reach: to give them reasons to support Prop 71.

And so we lumbered on, gaining friends, ignoring opponents, collecting signatures.

"Let's send a message," said Bob Klein, "Let the opposition know we are not going away. We need 565,000 valid signatures — let's collect a million!"

Sometimes it seemed like magic.

Once a bunch of us were on a bus, going up a steep San Francisco hill. In a terrifying instant, the bus lost traction, and started rolling backwards — but the driver just steered the rear of the bus onto somebody's lawn.

"Happens all the time," said the driver, turning the bus to try another way.

And once we had a rally, and walked across the Golden Gate Bridge, the blue ocean far below us…

How to raise money for the campaign? First, what could **not** be used? Bob made it a rule that the campaign would take no money from drug companies, or any biomed corporations.

Bob sold part of his company's assets, and took out loans on his house. Over the course of the campaign, he personally donated $3.4 million dollars. He leveraged that money, asking millionaires to match his donations — "If I chip in a million, will you do the same?" In the end, the cardboard thermometer was solid green — $34 million.

But against us was something which made me tremble a bit — the Catholic Church, the largest non-profit organization on earth.

Gloria was very Catholic, strong in her faith. But she had once gone against the church — and had seen what that could be like.

We were at a time in our lives when we could not afford any children, and Gloria was taking birth control pills. Uncertain about this, she confessed her "sin" to a priest. The words were scarcely out of her mouth when he came hurtling out of the confessional, shrieking at her, calling her a "Murderess!". It was a while before she went back to that church.

Bob raised money for TV ads here and there — but the church owned TV and radio stations, whole chains of them.

We used friends' living rooms and Bob's office as gathering places; the Catholic church had more than 100 churches in San Francisco alone.[5]

But you thought churches were not supposed to take part in politics, especially since they pay little or no taxes? That did not seem to slow them down.

At church, they passed out anti-stem cell research information, which Gloria brought home and showed me. At my grandson's Catholic school, specially trained priests visited, explaining the alleged evils of embryonic stem cell research.

The Republican Presidential Platform gave an official "Oppose" to embryonic stem cell research: a gift, it seemed to me, to the ultra-conservatives in their ranks.

Other communities of faith (like the Presbyterian church and the Judaic faiths) gave us their endorsements.

Most people liked us. They understood our motivation; we were fighting for our loved ones, something people understood.

The polls were scary. Generally, an initiative with a dollar cost starts out with a high approval rating, but then goes down as people consider how much it may cost.

Our initiative's polling started with just a two-point lead: 45% support, 43% oppose. We could not afford to lose a single vote.

Hollywood helped us! A motivated group of accomplished actors and other influencers added their voices in support.

Superstar Brad Pitt did a cheerful TV media opportunity, as did numerous other stars of stage and screen: like Rhea Perlman of CHEERS, and Dustin Hoffman who made autism understandable in RAINMAN, Jerry Zucker the director who made the world laugh in AIRPLANE, and cry in GHOST;

Arnold Schwarzenegger, star of many box office smashes like TERMINATOR, now Governor of California;

Sherry Lansing lent us her voice: CEO of Paramount Pictures, when it produced the movie TITANIC. Ms Lansing was the first woman to

---

[5] https://en.wikipedia.org/wiki/List_of_churches_in_the_Roman_Catholic_Archdiocese_of_San_Francisco

receive a gold star for motion picture production on the Hollywood walk of Fame.

Michael J. Fox's personal courage turned the suffering of Parkinson's disease into a potent political force for advocacy and fund-raising;

Almost literally with his last strength, Christopher "Superman" Reeve did a commercial for us, urging a YES! vote on Proposition 71: which his family insisted be run, even after our champion passed away.

And millions of people considered something new.

**We won** with a 59.05% YES count, against 40.95% NO — almost a 60–40 win. Seven million Californians trusted us with their votes.[6]

We had no way of knowing, of course, how many battles lay ahead.

---

[6] https://en.wikipedia.org/wiki/2004_California_Proposition_71#Results_of_vote

# 7 Sued by an Embryo?

Alameda County Judge Bonnie Lewman Sabraw (website) presided over the legality trial of the California Institute for Regenerative Medicine. Bob Klein makes a point of order.

"All rise,' said the bailiff. In a rustle of robes, the Honorable Judge Bonnie Lewman Sabraw entered her courtroom.

The California Institute for Regenerative Medicine, (CIRM), was under attack.

Apparently, it was not enough to win. You also had to defend your victory in court, no matter how clearly the will of the voters was expressed.

Lawsuits came at us from all directions. They might be frivolous (and in my opinion, they were) but they had to be defeated, before our stem cell program could receive bond funding to give out research grants. We were essentially frozen.

CHRISTOPHER REEVE & ROMAN REED
AT FUND RAISER 5/1/99

"You are more like Superman in real life than in the movies," said Roman Reed to Christopher Reeve.

Governor Arnold Schwarzenegger authorized a loan of $150 million to get CIRM started — and every penny was paid back with interest. (CIRM photo) On left, the ultimate stem cell research advocate, Bob Klein.

Against us were:

1. The toughest anti-tax group around, the National Tax Limitation Foundation (NTLF). This was the late Howard Jarvis group which had made it so difficult to raise taxes in California: even for the most important programs.
2. A group called People's Advocate was represented by Dana Cody, the woman who reportedly persuaded President George W. Bush to sign legislation (in the middle of the night, in his pyjamas!) concerning the Terri Schiavo case, where an apparently brain-dead woman was kept on life support for months in a right-to-life legal battle.
3. The California Family Bioethics Council was a just-formed group (reportedly the day before this first court date!) from the James Dobson "Focus on the Family" evangelical empire; it owned more than 60 religious radio stations, and was a force behind the Religious Right wing of the Republican party.
4. A fictitious embryo, "Mary Scott Doe", was represented by a group called the National Association for the Advancement of Pre-Born Children.

After due consideration, the group representing the embryo was denied standing, and all the other lawsuits were combined into one.

Both sides asked for a quick resolution of the case. Their side claimed the stem cell program was unconstitutional and should not be allowed. Our side felt their case was pure nonsense, and should be tossed out. Well, that's how I felt anyway.

But Judge Sabraw said the case was an important one, and should be heard.

We began.

The plaintiffs had many objections to the California stem cell program, but chief among them were:

1. Because state money could not go to institutions unless they were under State control, they tried to make our program sound like a rogue agency.
2. Initiatives are only allowed to cover one subject. Proposition 71 (according to the opposition) had several.
3. They felt our board of directors might abuse their positions to get grant money for their home organization.

When they talked, things looked grim; a flood of big words and lawyer-ese.

But we would have our turn.

If chief counsel James Harrison had been in charge of the cruise of the ship Titanic, it would have made a boring movie — because nobody would have died. The co-writer (Bob Klein being the primary author) of Proposition 71 would have foreseen the danger, avoided the icebergs — and brought everyone safely home.

Harrison's work, like that of Bill Lockyer, Attorney General of California, would be mostly behind the scenes.

The person who would actually speak for us, to make our presentation, was Tamar Pachter, Deputy Attorney General.

When I first saw her in the courtroom, I thought she was too young to be defending all the hopes and dreams of millions of suffering people and their families.

But once she began to talk, I mentally kicked off my shoes and leaned back. Ms. Pachter was clear-spoken, well-informed and tough. Her introduction (taken from the transcript) read like cold fire.

"We are here to make sure California voters get what the Constitution guarantees them…and what they voted for, by overwhelming majority.

"I don't doubt the sincerity of plaintiffs who are opposing Proposition 71 here. They have strongly held beliefs that merit consideration. But those beliefs did not prevail in the most important forum, the ballot box.

"The Constitution is the bedrock foundation of our democratic system; it must be preserved. The court's role… is not to use the Constitution to interfere with the initiative process, but to uphold it by… legitimate exercise of the voters' power.

"The implementation of Proposition 71…is fully compliant with all Constitutional and statutory requirements."

Then she went to work on the individual charges against us. I did not understand everything she said, but the opposition did. It got very quiet in the room.

In brief, her response to their three main attacks:

1. Far from being an institution out of control, the California stem cell program was now an official part of the State government, and subject to numerous state laws and controls.
2. "Single subject"? Every word of Proposition 71 had to do with stem cell research therapies' hunt for cures; it was absolutely "single subject".

3. All board members were subject to conflict of interest regulation and supervision… including routine public audits. If a grant might affect their institution, they were not allowed to vote on the grant or even discuss it.

The plaintiffs took an hour and a half to deliver their side. Ms. Pachter took 15 minutes to gut their positions.

When she was through, the opposition looked at each other. There was much reaching for the water pitcher.

But even if we were winning, (and I thought we were), as long as the lawsuits continued, our money supply was frozen — and with it, the research. All we had was the $3 million the state had guaranteed for the hiring of staff, nothing like the $300 million a year we had won to fund the research.

For most government programs, if their funding is cut off, they die.

But Bob Klein came up with something interesting. Now you may think I mention Bob's name too often. But it would be false not to.

Watch this:

In his previous life, Bob developed the California Housing Authority, to make low-cost loans for people of below average means. While there, he came across an obscure legal procedure called a Bond Anticipation Note, (BAN) and came up with a creative way to use it. To the best of my knowledge, no one had ever done it this way before.

People would be asked to loan the program money, to help it get started. But if the program died — the loan became a donation — and they would not be repaid.

Imagine going to a bank and asking for a loan, telling them you *might* repay it!

That was Bob's plan. And, there were far-seeing people willing to help the research move forward by participating in it.

Here they are, the 15 BANs purchasers:

Blum Family Partners
The Jacobs Family Trust
The Benificus Foundation
The Eli & Edythe L. Broad Foundation
The Moores Foundation
William K. Bowes Foundation
Mr. J. Taylor Crandall
Gordon and Betty Moore Foundation

Dr. Gordon E. Moore
Jewish Community Endowment Fund
H&S Investments I, LP
Seventh Street Warehouse Partnership
Steven L. Swig and Mary Green Swig
The David and Lucile Packard Foundation
The Sandler Family Supporting Foundation

Altogether, $45 million was raised with the BANs.

When it was clear we would win the lawsuits, Governor Arnold Schwarzenegger authorized a $150 million loan to CIRM, so we could get started.

At the press conference about this loan, Bob Klein spoke, and Arnold Schwarzenegger spoke, and then I had the privilege and honor of giving the patient-advocate response, to express our deepest thanks and appreciation.

I spoke for five minutes, but the only memorable part was the last sentence:

"The California stem cell program is the glory of a state, the pride of a nation, and a friend to all the world."

Not long after that, at the courtroom… we heard the judge's decision.

"The court finds that the plaintiffs have not shown that (Prop 71) is clearly, positively, and unmistakably unconstitutional. The Act (therefore, is) valid."

But — *the plaintiffs appealed.* The process dragged on and on, another two years.

Here is the Appeals Court Conclusion:

"DISPOSITION…Proposition 71 suffers from no constitutional or other legal infirmity. Accordingly, we shall affirm the (trial court's decision) upholding the initiative."[1]

The plaintiffs *appealed the case to the California State Supreme Court* — which also affirmed the decision.

And that was that. Every penny of the BANs purchases was repaid, with interest.

Now the stem cell battles could begin.

---

[1] https://www.cirm.ca.gov/about-cirm/newsroom/press-releases/02272007/california-stem-cell-project-prevails-appellate-court
Note: For a fuller explanation of the court cases, see "STEM CELL BATTLES," my first book on the subject.

# 8 Where I left My Heart

Karen Miner, Bob Klein, Gavin Newsom, Don Reed and other advocates unite in support of the California stem cell program. (picture source unknown)

We needed a central headquarters for the new stem cell research grant funding agency.

The first question was: where? Naturally, any city would want to have the California Institute for Regenerative Medicine (CIRM), bringing with it pride, jobs and money. Right now, the headquarters were in donated office space in Emeryville, but that was too small for long-term use by the agency.

**FIRST MAYORAL HANDSHAKE goes to Don Reed, whose son Roman suffered spinal injury was instrumental in formation of the the Reeve-Irvine Research Institute.**

When San Francisco was chosen as Head Quarters for CIRM, Mayor Newsom's first handshake was with a patient advocate Don Reed, next to smiling wife Gloria follow Gavin's arm... photo from the late lamented San Francisco Sentinel.

Bob Klein speaks on the future of CIRM; Mayor Newsom waits his turn. (Chronicle/Darryl Bush)

The top six contenders were: Los Angeles, Berkeley, Sacramento, San Francisco, San Jose and San Diego.

Each had reason to be considered for the honor. Los Angeles, San Diego and Berkeley had magnificent universities; San Diego was also home to clusters of biomedical companies; Sacramento was our state capitol and the hub of politics; San Jose was an energetic city on the rise; San Francisco was the home of pioneering researchers like Dr. Gail Martin of UCSF, who coined the expression "embryonic stem cell" — but only one city could be chosen.

The mayors would give presentations on why their city should be chosen. Some were pure lectures, others provided tours on buses, driven around the relevant city. When these happened, I went along, because it was a public meeting and mainly because I asked.

Right now, there were 2-3 board or committee meetings a week, and (with rare exceptions) I attended them all, a practice I continued for many years. As a patient advocate, CIRM was the biggest thing in my life — to touch the pulse of history?

One Mayor stood out: Gavin Newsom of San Francisco. He had unmistakable star quality, as if he was in technicolor.

He had an amazing memory. Gloria and I met him at a fund-raiser once, and talked just briefly. The next time, he remembered our names, and asked about Roman.

Most of the Mayors relied on notes for their bus ride tours, or checked with an assistant. Not Newsom: he just bounded aboard, sat behind the driver, gave a big smile, and started talking. He seemed genuinely glad to be with us, and talked a lot — - which is maybe why he always seemed to have a sore throat.

One of the reporters asked me: "You come to all these meetings — which city do you want?"

And I said: "I am glad I don't have to decide that, because every city is making a terrific case — but if I had to choose, I would go with that great song by Tony Bennett — "I left my heart in San Francisco!"

There was also the small matter of the laboratory buildings, which needed to be built throughout the state.

In time-lapse photography, a building can leap to the sky in seconds. In real-life, of course, that same building may take decades to dream, design, fund, get permits for, get workers, sign contracts — and may still not get built.

The need for the labs came clear to me on a walkthrough I did in a hallway at UC San Francisco.

"Stop for a moment," said the scientist-guide, "Notice the lab on your right?". I did, although not entirely sure what I was supposed to be looking at.

"Now, the laboratory on your left. What is the difference?"

I turned my head back and forth, staring hard.

"I can't see any difference," I had to admit.

"That's because there isn't any. These two labs were built identically — to satisfy the Bush regulations. One was built for embryonic stem cell (ESC) research, which California allows; the other is for federally-funded work, which could not, at that time, include ESC research.

"If even one dollar of federal funding was used for ESC, even a ballpoint pen — the university might lose all its federal funding.

'That's why California funding is so important — because the scientists need a place to do research without politics affecting it."

"Bricks and mortar" are expensive. Construction costs can devour funds, and short-change research.

But not here. Prop 71 was designed to prevent cost overruns. First, an absolute limit was set on how many dollars could be spent on lab construction: no more than 10% of the total project: $300 million out of the total $3 billion. We wanted labs that were solid, well-equipped and modern, but not wasteful.

We had no idea of the surprise that was waiting for us.

On October 28th, 2005, the Facilities Working Group took on the challenge. They would invite proposals from institutions around the state, then review them and prepare the information to put before the governing board, who would decide.

Let's listen in on an actual meeting, just a minute or two, to see the process in action. The meeting took 140 pages to transcribe; I cut it down to two.

BOB KLEIN: "Joan Samuelson was going to be able to attend. But Joan is — "

JOAN SAMUELSON: "Present!" (Joan has Parkinson's disease)

BOB KLEIN: "We are excited that Joan made it."

Committee Chairman Rusty Doms: "We have on the agenda, the Pledge of Allegiance, but since we don't have a flag yet, I think we unfortunately have to dispense with that today... On this committee we have four members from the real estate world, six patient advocates, and Bob Klein, our ICOC Chairman...I'm from Southern California, born and raised there. I've been in the real estate business basically all my life, including the development of two very large hospitals."

DAVID SERRANO-SEWELL: "I have the honor of representing the Multiple Sclerosis and ALS community as a patient advocate."

BOB KLEIN: "The Initiative provides ...$300 million for construction, hopefully with at least a one-to-one match." (IMPORTANT: he is talking about matching funds, from a different source.)

DAVID LICHTENGER: "I've been in construction...for over 20 years. In the last 5 or 6 years, I worked heavily on the life science side, doing a lot of labs, technical facilities, clean rooms...also some hospital and health care work as well."

ED KASHIAN: "My background is in real estate, universities, state colleges."

JANET WRIGHT: "I am a cardiologist (heart surgeon), one of the patient advocates on the ICOC (governing board). This group is about getting science a home — I am excited to participate."

JEFF SHEEHY: "I'm a patient advocate representing people with HIV and AIDS."

ZACH HALL: "I'm the President of CIRM right now. Most of my career was at UCSF. I was a neurologist once upon a time. Now I've become a science administrator."

JAMES HARRISON: "I'm James Harrison, outside counsel to the CIRM, and it's my job to help keep you all in line."

JANET WRIGHT: "A full-time job!"

BOB KLEIN: "James Harrison was one of the key attorneys who helped me with the drafting of Proposition 71, and is extraordinarily well-versed in the legal issues behind Prop 71, its structures and implementation. So we're very privileged to have his services."

CHAIRMAN RUSTY DOMS: "This is an extraordinary opportunity to turn hope into reality. People around this table have all been affected in some way or another, either themselves personally or a member of their family, with terrible diseases. I have a daughter who has serious learning disabilities."

ZACH HALL: "As you know, an important part of our mandate is to provide facilities that can be used particularly for human embryonic stem cell research.

Unfortunately, United States scientists have not been able to participate fully, because the major supporter of biomedical research in this country, the NIH (National Institutes of Health) has been restricted by Federal policies. Because of this uncertainty, most of the research buildings that institutions have…are not suitable for this very important work.

"What we envision would be having lab space with necessary basic equipment; that is, incubators, sterile hoods, freezers, microscopes.

"$100 a square foot (might be the price). But how much equipment do you put in these…do you (for example) have to have a separate mass spectroscopy facility?

"You might decide…to set up a core facility in which you would have a very expensive piece of equipment that would be shared, used by a lot of people…

"There are many questions to answer. What space and facilities do they presently have for human stem cell research? How large would the proposed new facility be? Do they plan to expand an existing facility? What is the budget and timetable for completion? Is this a reasonable cost? Are the timelines and milestones satisfactory? Is there adequate oversight? What is the level of institutional commitment?"

BOB KLEIN: What we have before us is a tremendous challenge in timing…facilities that can be built within two years of a grant award. We'll probably need to have a partnership with cities to accomplish that…such as fast-tracking projects through permit processes, etc. It is imperative we protect our institutions and our researchers from (political) pressures that may … come upon us in a blink of an eye…

"But it is with great optimism that I look at the group assembled here. I know they are up to the task."

DON REED: "On behalf of people who are suffering, thank you so much for the incredible amounts of work that will be demanded of you for almost no money.

"I have to tell a story about (committee member) Sherry Lansing since she's not here to defend herself...I am movie crazy. So, as you know, Sherry was the President of Paramount Motion Picture Studios. I had to ask her, how can you, the head of this powerful organization (she did the movie TITANIC) how can you justify the time to work on this particular assignment?

"And she said, "*It is the most important thing in my life*".

"I believe you all feel that way. Thank you!"

CHAIRMAN DOMS: "Thank you. Too bad there are not a lot more people like you who support us and help us with what we're trying to do here."

DON REED: "There ARE a lot. I am just noisy!" — -END OF TRANSCRIPT

And now, thanks to the miracle of time-lapse scriptography, (i.e. print), let's jump ahead, skipping thousands of hours of meetings and other people's hard work — and see what happened.

First, CIRM funded Shared Labs around the state in June 2007, so the research and the training of scientists new to stem cells could get underway as soon as possible.

Then the Facilities Working Group took on the larger task of distributing grants for major stem cell research facilities on campuses around the state. On May 7th, 2008, the governing board of CIRM voted to distribute $271 million to 12 institutions to build stem cell research facilities throughout California.

AND — remember the matching funds requirement, a way to get more bang for California's research buck?

The winning institutions had to bring along some money of their own — and they did — a lot!

California's $271 million for Major Facilities was leveraged into $1.15 BILLION.

Oh, and the location of the headquarters for the California Institute for Regenerative Medicine?

"I left my heart in San Francisco..."

# 9 Muotri's Robot

Alysson Muotri is locked in a battle with autism, which stole his son's ability to speak. (igm. ucsd.edu)

"Chronic diseases are defined... as conditions that last 1 year or more... they are the leading causes of death and disability in the United States... (as well as the) leading drivers of the nation's $3.8 trillion in annual health care costs."[1]

Every chronic disease is cruel; an illness that lasts "a year or more" — or which may never go away at all?

But some are worse than others; they attack children.

---

[1] https://www.cdc.gov/chronicdisease/about/index.htm

Alysson Muotri is a cheerful scientist with a Brazilian accent. When you talk to him, even on the telephone, you will enjoy the conversation. He always seems to be smiling.

He is also extraordinarily talented, "on the cutting edge of science", according to Rusty Gage, who persuaded Dr. Muotri to relocate from Brazil to San Diego.[2]

Recently he became involved in the effort against COVID-19, and doubtless made a strong contribution. But in my humble opinion, I hope he will not stay there.

A flood of scientists and doctors are already fighting the pandemic, and with appropriate levels of funding. That is right and proper.

But Dr. Muotri's son has autism.

I first became aware of that condition when my son Roman and I were in Southern California in 1994, at a rehab center, fighting his paralysis.

A frantic pounding on the apartment's front door got my attention. A frazzled-looking woman said:

"Tomorrow morning at ten, you will hear screaming from our backyard. Do not call the police, nobody is being abused. It is just my son has autism."

Then she ran off.

I thought no more about it, until next morning, when the screams began.

In the back yard next door, a little boy, all by himself, was calmly sitting in a sandbox, playing with a neatly-arranged row of toy trucks — and screaming...

In the years since, I have met half a dozen families with an autistic member, and while every one of their lives is different, none was made easy by the condition. One young adult liked to leap on top of his mother while she was asleep. He liked to do that when he was little, but now he was almost a full-grown man.

Each of these parents worries about who will look after their disabled child, after they are gone. One woman was reportedly in such emotional anguish when she could not find a program to help her son that she killed him, and then herself.[3]

---

[2]— "Autism researcher Alysson Muotri's audacious plans for brain organoids", by Hannah Furfaro, 12 August 2019

[3] https://www.mercurynews.com/2012/03/08/parents-of-autistic-children-speak-out-on-sunnyvale-murder-suicide/

Dr. Muotri's son Ivan is non-verbal, and his communication (based on a video I saw of the family) seems limited to facial expressions. He is a beautiful child, seeming confident in his parents' love. But still there are social, mental and physical problems — and he is non-verbal...he cannot speak.

Ivan now has seizures: ten to fifteen every day.

Perhaps one in a hundred children have some form of autism, including Rett Syndrome, Angleman's, Asperger's, and others.

Can we cure them? Not yet. We cannot even come close, while one obstacle is in the way.

There is *no accurate model* of a human brain on which a scientist can experiment. We cannot say to a person, "Excuse me, can we open your skull, and dig around in your brain for a while?"

Fortunately, there may be an alternative.

Imagine a little robot, shaped like a metal spider. It is cute, non-threatening, and it walks, (not too well), occasionally bumping into walls.

It is controlled (again, not too well yet) by a patch of cells in a dish of salt water, which is itself influenced by a computer. The patch of cells is called an organoid. It is almost invisibly small, and has been called a "mini-brain".

This robot-organoid-computer connection is by no means a human being: not even close. It is a speck of tissue, not a person, no more than a dot of cheese is a pizza.

But it may be a step in the direction of cure.

When Dr. Muotri coordinates messages between computer, organoid, and robot, he is modeling the brain-nerve-body connection. The cells in salt water can be experimented on, to see what makes them work, and what does not.

This ingenious tool may help find a cure, not only for autism, but for many diseases of the brain and nervous system.

Dr. Muotri is a hard-working scientist, no question. But he also has a flair for color and spectacle, as when he sent organoids to outer space, to test them in zero gravity.

And he has studied the DNA of Neanderthals, our cave-dwelling ancestors.

But in the patient advocate community, I suspect he will always be remembered for something small: a tooth.

First, the problem. How do you take a tissue sample from a person who cannot be touched? Some autistic folks cannot bear to have someone hold them, even gently — let alone poke a needle into their arm, to remove a piece of flesh.

But as we all know, leaving a tooth under the pillow brings the tooth fairy... so the kid feels his tooth is loose, he wiggles it out, leaves it in the proper place, in exchange for a dime or a dollar or stocks and bonds — and that was how Dr. Muotri got all the tissue samples he needed...

In the lab, Muotri is using the brain model to learn how nerve networks from autistic folks are formed. The communication between cells with autistic genomes is different from those of healthy individuals.

If the autism is caused by a (mutating) gene, Muotri may use enzymes to fix the mutation, or insert a correct copy of the gene. In some cases, the genetic information is not enough, and he must screen the organoids to search for (new drug) treatments. This approach has already paid off: he has been able to find novel treatments for rare syndromes.[4]

He has the creativity and drive. If left alone politically, and provided with adequate and reliable funding, Alysson Muotri may find a cure for autism.

He — and his son — are why we fight: one of the many human reasons why the California stem cell program must go on.

---

[4]—Alysson Muotri, personal communication

# 10 The Loss of a Friend, and Two Shark Stories

Vanessa Lupian and her Mom. Vanessa survived a liver transplant. (blog.cirm.ca.gov)

If you have been following my small series of books, you might remember the heroic efforts of Will and Mary. Eternally cheerful Mary had a disease of liver failure; Will had a love for her as big as all outdoors.

For years they drove back and forth across the United States, in search of a transplant, a compatible liver from someone who had recently died.

They never found it. A week ago, their struggle came to an end. At Mary's request, they drove back to Will's childhood home, where they had spent many comfortable years; and there she passed away.

I have known Will since we were kids; his pain is beyond any words of mine.

Author scrubs the floor of Marine World aquarium, as a sevengill shark cruises by. One third of a shark's bodyweight is its liver. (Photo by Dave DiFiore.)

What could sharks have to do with such a tragedy? Perhaps nothing more than an attempt to smile on a cloudy day.

I was a professional scuba diver at Marine World Africa USA, from 1972-86. I swam with sharks hundreds of times, and still have all my limbs. [1] In a moment I will tell you two shark liver stories.

But first, imagine if your own liver did not work. What would your life be like?

For me, it would be short. Even as an adult, I find it difficult to control my food intake, which is why I weigh 220, instead of 190, which would be appropriate.

But if I was a child? That would be still worse because I might not understand the need for food restriction.

Do you remember Vanessa Lupian, mentioned in my previous book "California Cures", on page 67?

She was six years old and stricken with liver disease; it was awful. Her family, fighting to keep her alive, carefully monitored every gram of her

---

[1] Notes from an Underwater Zoo, Don C. Reed, Dial Press, 1981

food intake. She would sometimes be so hungry, having eaten her small portion, that she would ask to just smell someone else's food...

Want a look at the kind of dietary restrictions a person with a liver condition has to endure? Here is a partial list of suggestions from the Fatty Liver Foundation.[2]

None of that even sounds like food! Brussell sprouts? What crime did I commit to have to eat that — and not even very much!

It was agony for Vanessa's family as well, having to deny her food when she was so hungry. But if she went off the diet, it could send her to the hospital, or worse.

When her Mom found out a liver transplant could "change everything, the risks frightened her."...But when she told her daughter, Vanessa burst into tears... "Mom, this is the happiest day of my life!" she said.

At age 10, after getting a new liver, a little girl had her first French fry..."[3]

So why doesn't everybody just get a new liver, if they need one?

First, as in the story of Will and Mary, there is a shortage of donated livers.

Second, to the person who receives them, transplanted organs are a foreign body. He/she may need to be on immune suppression drugs for life — and these bring their own set of medical problems, some very serious.

Third, a liver is complicated! According to JohnsHopkinsMedicine.org, the liver does 500 subtle and different chores every day, and we never know it — until something goes wrong.

Now, the mouse and the scientist.

Wilger Hollenbring, M.D., Ph.D., works at the Liver Center at the University of California at San Francisco. (I have quotes from from his documents at CIRM — mistakes of interpretation are of course my own.)

As for Dr. Hollenbring's mouse, it had liver disease. Could it be made well?

As a stem cell enthusiast, my first thought was: grow a new liver from the animal's own cells. Take out the old liver, put in the new. It should not be rejected, being from the original owner.

---

[2] https://blog.cirm.ca.gov/2012/09/17/liver-transplant-saves-girls-life-but/
[3] https://www.fattyliverfoundation.org/nash_diet

Not being a scientist, I do not know the problems they face: it takes time to grow cells in a dish of salt water — time a person with liver failure may not have.

But what if the needed cells could be grown inside the patient? The disease caused scars on the liver; perhaps the scars themselves could be reprogrammed to become new and useful tissue.[4]

That is one of the things Dr. Willenbring and his team accomplished — and while there are of course many tests ahead, the results are promising. As he put it:

"We have made significant progress toward our goal of generating, in the laboratory, human liver cells that are therapeutically effective in mouse models of human liver failure…"

Because of such work, it may become possible to save the lives of some of the 35,000 Americans (children and adults) who die every year of liver failure. Think of that number — if the average Junior high school has 803 students, those deceased are roughly the population of 44 middle schools.

But I promised you two shark stories?

The first involved a shark which died for no apparent reason. I found it on the tank floor: a sevengill shark, about four feet long, beautiful in its own way: golden eyes which could suck back into its skull if in danger; a brown back with clusters of black spots, as if a leopard had dipped its paws in ink and then walked on the shark.

The vet put the dead shark on its back on a cutting board, , took a scalpel and swooshed it down the belly.

Something enormous and grey-pink burst out.

"That's the liver," said the vet. It was huge, about a third of the animal's weight, and full of oil, so much that sharks supported an industry for a while — you have heard of cod liver oil? Some of that was shark liver oil…

We never knew for sure why that shark died, but it was Summer, and the water temperature may have risen beyond the shark's ability to handle change.[5]

The second story is not my own, so I will not name the aquarium where it happened, except to say it was NOT at the Oklahoma Aquarium, which has a magnificent display of bull sharks, very dangerous animals,

---

[4] https://www.cirm.ca.gov/our-progress/people/holger-willenbring
[5] "Sevengill: the Shark and Me". Don C. Reed, Sierra Club Books, 1986

but with which the aquarium divers routinely swim. They know what they are doing.

But in the aquarium which shall be nameless, a massive-bodied bull shark was causing problems, becoming increasingly aggressive toward the divers who cleaned its tank. The situation got worse and worse until finally (on a day when the park was closed) a diver took a speargun and ended the problem.

The shark's death was listed as "liver failure", which was technically accurate — that being where they shot the shark.

# 11 Killer in My Classroom

Christina Chatzi (ohsu.edu) brings youthful enthusiasm to the fight against chronic disease; Rusty Gage, President of Salk Institute (salk.edu) is a long-term veteran — CIRM unites their energy.

Call him X, because so much about him is unknown.

In the years I was acquainted with him, his parents were not a part of his life; I don't know why. But his grandparents were. And I knew them, somewhat.

X was athletic and played on some of the same teams Roman did; it was natural for us to give him rides to and from practice. He seemed quiet and normal; nothing stood out to make him seem different: someone who might kill.

He was assigned to my 8<sup>th</sup> grade English class, where he kept up, grade-wise. He was no problem behaviorally, and every morning when I stood in the doorway and said "Good morning!" to each of my students (everyone deserves a greeting) he would say "Good morning!" back.

Now: in my class, everybody did an essay every week, with the exception of book report week, (once a month). That meant you had no choice but to learn how to write. One high school teacher said she could spot one of my kids right away, because they were not afraid to write.

But one day X wrote an essay about his English teacher, who used to work at Marine World, and who was the ugliest person he had ever known. It was not difficult to figure out his teacher's identity.

The essay went on and on, one insult after another. I had never seen anything like it. So I took it to the Principal, who called the grandparents in. He told them he counted seventeen insults on the paper, and if I wanted to press charges (!) there could be major fines, as much as one thousand dollars per insult.

So X was called in. He apologized, and of course I let it go. But I did not understand the reason for these spectacular insults. On the way back to class, I asked him: had I ever been mean to him? Or unfair? He said no. So... why?

He shrugged. Over the years I came to believe he meant it as a joke. He did not seem to realize he had stepped across any sort of line. So, we went on, as before.

After he graduated, I heard of him, now and again, several times over the years. I cannot vouch for the details of these events, except for their horrific outcome, which is, unfortunately, a matter of public record.

First, I read a magazine article, where X apparently tried to commit suicide, climbing over a bridge barrier. He was rescued by a kindly couple, who took him under their wing, practically adopting him.

But X got into selling drugs, and perhaps he used his own product. He seemed to have become paranoid, as if his companions had turned against him.

He took three friends into the hills, supposedly on a picnic.

There were two other boys, and one girl. X took out a shotgun. He killed one of the boys outright. Then he shot the girl in the stomach, and she screamed and said "I'll do anything, just please don't kill me!". X turned to the other boy, frozen with fear (doubtless thinking he was next), and said, "Never leave a job half-done," and reloaded. He shot the girl again, killing her.

Then he handed the shotgun to the survivor, saying he "wouldn't need that any more", got into his car and drove off. He saw a policeman, made an obscene gesture at him, then roared away. A high-speed chase ensued.

X turned a corner at high speed, crashing his car, flipping it onto its roof. But he survived.

He was tried and convicted, sentenced to die, put on Death Row.

I will believe till my dying day what I told his lawyer in a telephone conference: that X was damaged; his brain was just not right.

The fact that he killed two people cannot be erased. Their families' lives are forever ruined.

But his life too was broken, and from early on. I cannot help but wonder: was there something physically wrong with his brain? And if so, could it have been spotted earlier, and perhaps repaired?

He died (of "undetermined causes") while on death row, San Quentin, April 20th, 2021.[1]

Did he have a mental illness, perhaps a form of schizophrenia?

Schizophrenia afflicts 1.1% of the American population, roughly 3 million people. The symptoms are hallucinations (hearing voices, seeing things that are not there) — and sometimes deep feelings of persecution.

Violence? "Thankfully, violent behavior is comparatively rare. Most people with schizophrenia are *never violent* and do not display any dangerous behaviour. However, a small number do become violent when they are suffering from the acute... psychosis because of the hallucinations and delusions... tragically, ... many people with schizophrenia succeed in killing themselves."[2]

Life-long drug treatment can help keep symptoms in check, but there is no cure — and as many as ten per cent surrender to the disease by taking their own lives.

Is there a cure for this nightmare?

While working at Sanford-Burnham, Dr. Christina Chatzi asked herself: If one kind of nerve cell (neuron) caused a mental illness, could a second kind inhibit it?

"Knowing that "some inhibitory neurons rely on ...retinoic acid (a form of Vitamin A) ...Dr. Chatzi wondered if exposing embryonic stem cells to retinoic acid might result in these inhibitory neurons.[3]

Working with Greg Duester in his lab, Chatzi found that mice which could not make the retinoic acid in their bodies had a "serious deficiency

[1] https://patch.com/california/sanrafael/condemned-san-quentin-inmate-dies
[2] https://livingwithschizophreniauk.org/information-sheets/schizophrenia-and-dangerous-behaviour/
[3] https://www.cirm.ca.gov/blog/04122011/making-neurons-lose-their-inhibitions

in (certain) neurons...that deficiency has been associated with... neurological disorders, including schizophrenia..." In other words, the lack of retinoic acid in the body might bring on mental problems.

One of America's top neurologists is Dr. Rusty Gage. He is fighting the condition from an entirely different angle — not the nerve cells themselves, but the spaces and connections between them.

Stanford's Marius Wernig points to a protein which might be deadly. If Myt1l (mighty-one L) has mutations, there will be problems.

'Mytl1 mutations have been...found in people with autism, schizophrenia, major depression, and low I.Q..."[4]

Maybe in years to come, newborns will be tested for potential mental illness, and (if the parents wish) be treated for it before they leave the hospital...

All I know for sure is that the tragedy of X ruined three young people, as well as devastating their families.

Schizophrenia has no place in anybody's life.

---

[4] https://blog.cirm.ca.gov/2017/04/05/cirm-funded-team-uncovers-novel-function-linked-to-autism-and-schizophrenia/

# 12 Dr. Larry vs. Alzheimer's

The scientist most likely to defeat Alzheimer's disease? I would choose Larry Goldstein. (biology.ucsd.edu)

"The global epidemic of Alzheimer's disease is worsening, and no approved treatment can revert or arrest progression of this disease." — L. Goldstein, et al.[1]

In the early 1960's, I was a mid-level competitor in the sport of Olympic weightlifting. I worked at the York Barbell Club, the center of American lifting back then. Since everybody in the sport came back to

---

[1] https://pubmed.ncbi.nlm.nih.gov/31780819/

York, at least to visit, and maybe train on a Saturday at noon, I had the chance to meet America's greatest weightlifters.

The champions I had the privilege of knowing now were equally great — but athletes of the mind.

Like Larry Goldstein, whose life is dedicated to fighting Alzheimer's disease.

Hold that thought.

Do you remember the classic comedy *Splash*, where an old woman comes to work oddly dressed, with her bra on the outside of her sweater? It was just a funny moment, never explained. But if the movie had followed her home, it would have ceased to be a comedy. Forgetting the order of putting on clothes can be a sign of Alzheimer's disease (AD).

She might once have been a trusted businesswoman. But now, her career was over. She could no longer write checks, because she might pay the bills two or three times, or not at all. At home, she might forget who her husband was, or her children, or her own name. At night, she might need to be locked into her bedroom, lest she wander away.

There is no cure for the disease, and her only escape would be the end of life.[2]

<u>Every 65 seconds</u>, another American is diagnosed with Alzheimer's. In America alone, nearly six million men and women currently suffer the disease. It is the 6th leading cause of death.[3]

The mother of Bob Klein passed from Alzheimer's. "The worst part was when she could no longer recognize her own family", said Bob.

The cost? National Institutes of Health (NIH) estimates run well over $200 billion a year in healthcare for those living with Alzheimer's.[4]

And that is just money — how do you put a price on suffering?

"Unless you've experienced Alzheimer's first-hand, it's hard to comprehend the toll it takes on an entire family. Watching a loved one transform from a strong, independent person to a shell of their former self is nothing less than devastating," said Lauren Miller Rogen, CIRM Alzheimer's board member and founder of Hilarity for Charity (HFC) the

---

[2] https://blog.cirm.ca.gov/tag/dr-larry-goldstein/ — "Stem Cell Battles", book by Don C. Reed

[3] https://www.alz.org/alzheimers-dementia/facts-figures

[4] —https://www.brightfocus.org/alzheimers/article/alzheimers-disease-facts-figures

non-profit that put out a TV production and puts on other events to raise funds for Alzheimer's research and advocacy.

"My mother taught first-grade for 35 years. She spent the last 15 years of her life battling Alzheimer's, and the final six in a bed unable to speak, feed, or care for herself. Fortunately, my family was among the fortunate few who could afford quality care for her. For so many, proper care is just financially impossible."

LMR — personal communication.

Financially, the worst is yet to come.

"By 2050, Alzheimer's is projected to cost… upwards of one <u>trillion</u> dollars a year, crippling America's healthcare system… "[5]

But the California Institute for Regeneration (CIRM, the California stem cell program) is fighting back.

Let me introduce a "top ten" of the CIRM scientists challenging Alzheimer's, and then we'll get back to Dr. Larry.

At Gladstone Institute, **Yadong Huang** has been working with an important kind of nerve cells (*GABAergic* neurons), which regulate chemical messages through the nervous system. Too much activity? GABAergic neurons calm it down.

The part of the brain that contains those cells is called the *hippocampus*. Damage there may cause AD. But could a transplantation of stem cells make things right?

That is what Dr. Huang is trying to find out.[6]

A completely different approach is being taken by Dr. **David Schubert**, of the Salk Institute for Biological Studies. Dr. Schubert reasons that aging itself may be a cause of Alzheimer's, as well as Huntington's disease and ALS.

Using J1-47, a derivative of *curcumin*, the active ingredient in turmeric, Dr. Schubert was able to reverse some of the aging symptoms (including memory!) in mice that were bred to age rapidly. Dr. Schubert's ingredient, J1-47, is now undergoing phase 1 clinical trials.[7]

---

[5] —https://www.cirm.ca.gov/our-progress/awards/human-ipsc-derived-gabaergic-progenitors-alzheimer%E2%80%99s-disease-treatment

[6] https://www.cirm.ca.gov/our-progress/awards/human-stem-cell-based-development-potent-alzheimer%E2%80%99s-drug-candidate

[7] https://www.cirm.ca.gov/about-cirm/publications/selecting-neurogenic-potential-alternative-alzheimers-disease-drug-discovery

He also is working on another possible game-changer, CAD-31.[8]

In collaboration with Victoria, Australia, UC Irvine's **Frank LaFerla** has used a virus to transplant neural stem cells, or growth factors, (food for the brain cells) into the brains of mice, to replace cells stolen by the disease.[9]

**Janet Baulch**, also of UC Irvine, used *exosomes*, tiny biological bubbles which transport genetic information to other cells; could these ease Alzheimer's?[10]

**Douglas Ethell**, at UC Riverside, is testing whether "immunotherapy… (may) provide cognitive benefits in a mouse model of AD";[11]

**Alexandra Capela** of Stem Cells, Inc, has worked to *restore* the lost memory cells: "Evidence from previous animal studies shows that transplanting human neural stem cells into the hippocampus improves memory, possibly by providing growth factors that protect neurons from degeneration…"[12]

**Roberta Brinton** of the University of Southern California, has tested a naturally occurring *steroid*, Allopregnanolone, for prevention and treatment of AD; how wonderful that might be, to prevent AD before it starts![13]

UC San Diego's **James Brewer** is comparing skin samples (biopsies) from AD patients and the "healthy elderly", to determine what goes wrong and why;[14]

**Tony Wyss-Coray,** of the Palo Alto Veterans Institute for Research, is studying how protein factors affect the aging nervous system; how

---

[8] https://www.cirm.ca.gov/our-progress/awards/neural-stem-cells-developmental-candidate-treat-alzheimer-disease

[9] —https://www.cirm.ca.gov/our-progress/awards/exosome-based-translational-strategy-mitigate-alzheimer%E2%80%99s-disease-neuropathology

[10] —https://www.cirm.ca.gov/our-progress/awards/es-derived-cells-treatment-alzheimers-disease-0

[11] —https://www.cirm.ca.gov/our-progress/awards/restoration-memory-alzheimer%E2%80%99s-disease-new-paradigm-using-neural-stem-cell

[12] —https://www.cirm.ca.gov/our-progress/awards/cirm-disease-team-develop-allopregnanolone-prevention-and-treatment-alzheimers

[13] —https://www.cirm.ca.gov/our-progress/awards/collection-skin-biopsies-prepare-fibroblasts-patients-alzheimers-disease-and

[14] —https://www.cirm.ca.gov/our-progress/people/tony-wyss-coray

appropriate it would be if some of our oldest warriors helped win the fight against aging disorders![15]

AD has a genetic component — but which gene or combination? **Mathew Blurton-Jones** of UC Irvine is examining what may be the Alzheimer's gene: CD33;[16]

Each of these scientists deserves their own article — or book — about his/her work.

But if I had to choose one scientist to sum up the struggle against Alzheimer's, it would be UC San Diego's **Larry Goldstein**, for many reasons.

First, a seeming digression. While Goldstein's papers are rich with the vocabulary of science, appropriate for the scientific audience which will read them, Dr. G. can also bring the science down to earth.

Goldstein can "talk people." Asked a scientific question, he can answer in words of one syllable. He has a sense of humor, and is fun to hear. This makes complex scientific ideas digestible by an average person like myself; ideas and results are only impactful when they can be understood by many.

That is why Barack Obama speeches have a 9th grade vocabulary level.[17]

If a scientist speaks at the 13th grade level (Freshman in college) or higher, he/she may be understood by their co-workers — but what about the rest of us? It is hard to support what we do not understand. If a scientist who wants public money, he or she had best talk small. Listen to Goldstein, see how it's done.

The sheer volume of work the man has accomplished is staggering. By my count he has published over 130 scientific papers on the most incredible subjects — he studied the internal workings of house flies (Drosophila) for more than 20 years!

He can be found where the scientific action is, generally in a position of leadership; after early research at Harvard, he became an investigator at the presitgious Howard Hughes Medical institute (1993–2012),

---

[15] —https://www.cirm.ca.gov/our-progress/awards/optimizing-differentiation-and-expansion-microglial-progenitors-human

[16] —https://languagemonitor.com/global-english/obamas-acceptance-speech-at-9th-grade-level/

[17] https://www.cirm.ca.gov/our-progress/awards/identifying-drugs-alzheimers-disease-human-neurons-made-human-ips-cells

concurrently with a professorship at the University of California at San Diego (UCSD). In 2006 he became Founding Director of the UCSD Stem Cell Program, a position he occupied for ten years. Since 2012 he has been Scientific Director for the Sanford Consortium for Regenerative Medicine- he recently became a member of CIRM's Board of Directors — and that is a very skimpy outline for a massive ongoing career!

He may work with the broadest of challenges, like this:

**Identifying Drugs for Alzheimer's Disease with Human Neurons Made From Human IPS cells**[18]

There are about as many possibilities for a project like that as there are bugs in an Amazon rain forest!

He may also get down to the most highly specific: "Amyloid-beta — independent regulators of tau pathology in Alzheimer disease";[19]

What does that mean? On the surface of an Alzheimer's patient's brain, are plaques and tangles. Previously, scientists thought these were interfering with the brain's function, thereby causing AD — and a cure might be found by removing them — but what if they were *a result, not a cause*?

A root cause of AD may be an inability of the brain to process cholesterol — which could turn the battle against the disease in an entirely new direction.[20]

Such research — and such scientists — are the reason the California stem cell program must continue.

P.S. For more on Dr. Larry, including his recent election to the National Academy of Science, read Kevin McCormack's excellent "A True Hall of Fame Winner".

---

[18] https://blog.cirm.ca.gov/tag/dr-larry-goldstein/
[19] —https://www.nature.com/articles/s41583-019-0240-3
[20] https://www.benefunder.com/health-wellness-causes/lawrence-goldstein/stem-cells-for-neurological-diseases

# 13 Making a Hemo Substitute?

CIRM scientist Tanishtha Reya might make emergency substitute blood. (stemcellpodcast. com)

Nearly six million people die every year from accident or gun shot.[1]

Of those deaths, an estimated one third could have been prevented if there was an adequate supply of replacement blood.

In this rich country, our available blood supply is called "alarmingly low".[2]

There just isn't enough blood to go around. Today, I will call 1-800-RED CROSS and set up an appointment to give some blood.

I made the call, they set up an appointment for me at a donation center, and I drove over. They asked me some questions, told me my heart skipped several beats in a minute (so I should tell my doctor about that) and slipped a needle into my arm.

---

[1] https://www.who.int/violence_injury_prevention/key_facts/en/
[2] CIRM Stem Cell Agency, May 25, 2017

I am told that one person's donation can save three lives.

But what if, instead of hoping for volunteers (I was *the only donor* in the room when I went in) hospitals made their own emergency blood supply?

Instead of borrowing blood from others, we might make it — from stem cells?

Enter CIRM.

Early in her career, Dr. Tanishtha Reya of UC San Diego wanted to take on that challenge: to make replacement blood cells, to have them available for emergencies. It would be great — if the funding could be found.

One of the most interesting grants given by the California stem cell program is the "Inception" award. This award "provides seed funding for great ideas that have the potential to impact human stem cell research, but need some initial support... to test their ideas, and give them the data they need to compete for more substantial funding."[2]

Dr. Reya applied for an Inception Award, $223,200, and focused on developing a stem cell-derived source of red blood cells.

Her grant application read:

"The goal of the proposal is to develop a human universal donor cell line... to produce functional red blood cells when needed...providing an unlimited supply of red blood cells for transfusion.

"The rising world-wide shortage in blood supply has highlighted the need for a safe, unrestricted source of human blood cells.

"Many factors have contributed to this crisis... medical procedures, such as chemotherapy, organ transplants and heart surgeries that often require blood transfusions are... increasingly common, the aging population is living longer and, as a result, undergoing more procedures that involve blood transfusions...."

The Chairman of the Board of CIRM would later say:

"Every two *seconds* someone in the U.S. needs a blood transfusion. But sometimes, due to a shortage of donors, there is not enough blood... and surgeries...have to be canceled.

"Creating a safe, unlimited supply of universal donor blood cells could have a huge impact, not just in the U.S., but world-wide," said Jonathan Thomas PhD, J.D., Chair of the CIRM board of directors."

The need was clear — she got the grant — and then?

Her "Research Objective" was "to create a universal...blood cell line that can be used to produce human red blood cells for transplantation."[3]

A universal blood cell line, that could be used by anyone, perhaps ending the world-wide shortage of donated blood... What happened?

"We have made significant progress toward this goal during the course of our work: identifying a powerful signal that can drive generation of blood cells...

"Further, we have defined conditions to promote growth of these cells for extended periods of time, and have developed strategies to trigger their development towards functional red blood cells...

"Our long-term goal is to advance these studies... so that the methods can be used to generate a safe and effective source of human red blood cells."

Dr. Reya's work with blood was years ago. Since 2011, the majority of Dr. Reya's work has been the fight against pancreatic cancer. Naturally, I am delighted that her strength is aimed at this terrible disease, with which I would one day become all too familiar.

But at the same time, she has a terrific idea on a universal blood source, and the world-wide shortage of blood continues.[4]

I spoke with her briefly, and she looked over this piece; with the enthusiasm of youth, she intends to go forward with both projects — if she can find the funding.

Tannishtha Reya: a name to remember, and support.

---

[3] https://www.cirm.ca.gov/our-progress/awards/reprogramming-human-stem-cells-blood-cell-generation
[4] https://www.usatoday.com/story/news/nation/2020/03/17/coronavirus-outbreak-us-faces-severe-blood-shortage-donations-plummet/5067055002/

# 14 Tarzan and the Obesity Demon

Muscular actor Gordon Scott played Tarzan and Hercules in many movies. (Star Stills photos, eBay.com)

I met Tarzan's daughter in high school.

I was rushing down the hallway, late for class, when suddenly a forearm shot out across my chest, stopping me cold, like being clotheslined in a football game.

"I understand you like Tarzan", said an attractive young lady, by way of introduction.

I admitted as much, when I got my breath back.

Navneet Matharu may have the gene/stem cell answer to obesity. (innovativegenomics. com)

Edgar Rice Burroughs' Tarzan of the Apes was a big part of my life growing up. I might be scrawny and weak, but Tarzan was physical perfection. I loved to see him do all the things I could not, swinging through the trees, flinging bad guys right and left. I devoured all 28 Tarzan books and saw nearly every one of the Tarzan movies — Elmo Lincoln, Johnny Weismuller, Jock Mahoney, Mike Henry, more — and my favorite, weightlifter Gordon Scott.

Gordon Scott in his prime was incredible: huge but not fat, chiseled. He could lift 300 pounds over his head, which was a lot back then. When his contract playing Tarzan ran out (six movies, two of which, "Tarzan the Magnificent" and "Tarzan's Greatest Adventure" were quite wonderful), he went to Italy and made Hercules movies. And when he fought in them (especially "Goliath and the Vampires") he made the impossible seem real.

"Gordon Scott is my father," said the young lady, "His real name is Gordon M. (Merrill) Werschkul." She later showed me a photo of her famous Dad in a tuxedo, autographed to her, Judy Werschkul.

Judy herself was healthy and strong (definitely strong!); clearly some of her father's genetic makeup had been passed down to her.

So here is a question: would you rather be lean and muscular like Tarzan (or the feminine equivalent) — or gain a couple hundred pounds and be obese?

This is not a trick question. Health is not a fashion statement. My wife Gloria was always beautiful to me — but then I saw her through the eyes of love.

Gloria suffered greatly from obesity. Her knees gave her terrible pain; walking became a hunt for the nearest place to sit down. And this was a woman who once could out-walk her husband; I practically had to run to keep up.

Fat also accumulates inside the body. It is said that every inch of blubber on the outside of your belly is matched by an inch of fat inside — around your heart.

She was not alone in her condition. According to the Center for Disease Control and Prevention, (CDC) nearly one American in two (42.4%) is clinically obese.[1]

Some of that problem we can control, with diet and exercise. When I was a kid, a middle class family had one car, so the kids walked everywhere. Also, there were no fast foods, so we could not stack on the blubber so easily.

But sometimes it felt like there was a fat demon, creeping up on us while we slept, making evil magic so we gained weight — while others did not.

In an earlier book, I had a chapter called "WHAT COLOR IS YOUR FAT?" based on Shingo Kajimura's work with brown fat. He has since moved his lab to the East Coast and we have lost touch.

But from him I learned there are three colors of fat: brown, which burns off as energy, so its lucky owner is lean and athletic; white, which stores energy as blubber, the "bad kind" of fat; and beige, which might possibly be changed to the "good kind", the brown fat.

A person with similar ideas is Dr. Navneet Matharu, author of a paper which might literally change the world.[2]

What if you could lose weight, without endless exercise or semi-starvation?

Now of course I am in favor of exercise. At 76 I still do martial arts every day, plus floor exercise and walking, plus baby-size weights —

---

[1] https://www.cirm.ca.gov/about-cirm/publications/crispr-mediated-activation-promoter-or-enhancer-rescues-obesity-caused
[2] "CRISPR-mediated activation of a promoter or enhancer rescues obesity caused by haplo-insufficiency", published in Science, 2019. Primary Investigators (PI) Christian Vaisse and Nadav Ahituv.

30–45 minutes a day. And I grudgingly admit we must eat right. My daughter Desiree lectures me endlessly about the values of vegetables, which I admire, from a safe distance.

At the doctor's office recently, I weighed 221. Aside from normal wear-and-tear of aging, I am fit.

But if I could lose ten pounds without effort — would I want that?

"In the vernacular of the peasantry" (as the Wizard of Oz once said) "Oh, Hell, yes!" ( Well, okay, he didn't say that part.)

Here is how it works with mice — and might one day work for us.

Think of two microscopic chains, side by side — our genes. "One from Mom, one from Dad", as Dr. Matharu said.

There are specific gene chains for obesity control. If intact, you are less likely to become obese. But if one link is broken, the obesity control gene will not work.

And listen to this. One of the reasons healthy people control their weight successfully is that after they eat, *they feel full.* They get full earlier, and do not feel the need to keep eating.

Obese people may <u>never</u> have that feeling of "satiety"; they do not feel full, but are always hungry, no matter how much they eat. That's tragic — and not their fault!

The broken-chain problem is called haplo-in-sufficien-cy. (The dashes were added to make it easier to say). Haploinsuffiency may bring on a host of diseases, including diabetes, epilepsy, autism, various types of cancer, kidney failure, heart disease, cardiovascular problems — but let's stick with obesity for right now.

What if you could fix that broken gene chain?

It can already be done with mice. You can take a normal mouse and make it fat — or take a fat mouse and make it normal — by altering the obesity gene.

Below is my interpretation of Dr. Matharu's paper.

Remember those double chains?

One double-chain gene is called Mc4R; another is Sim1. They help you keep your weight under control. But if one is broken...

Think about mice. If they have the broken chain, they become obese.

"But if they are injected in the hypothalamus (part of the brain) with something that can squeeze out more copies of the mRNA from a normal chain gene, this will increase the amount of protein being made — and the fat mouse becomes lean. It can puff up with blubber, or shrink down to normal, all depending on its genes."

But to drill a hole into the skull, to gain access to the brain? Surely that is too risky for human use?

It is being done right now, for sufferers of Parkinson's Disease. Called Deep Brain Stimulation, (DBS) it is a carefully guided cut into the brain, approved by the FDA in 2002.[3]

Naturally there are years of testing ahead. But the idea seems worth pursuing.

Not everyone wants to be built like Tarzan (or Jane!) — but we all should be healthy and fit.

If someone could make this science "real", turning theory into therapy, so we could have an operation and lose weight with little effort, he or she will be one incredibly busy person — and rich.

In America alone, there would be over one hundred million potential customers…

---

[3] https://www.cdc.gov/nchs/products/databriefs/db360.htm
[4.] https://www.ncbi.nlm.nih.gov/pmc/articles/PMC3785222/

# 15 *Before* the Lightning Strikes

At the 1975 Western Open, PGA champion golfer Lee Trevino was sitting under a tree in a rainstorm — and was struck by a bolt of lightning. He received burns on the back and a spinal cord injury. Fortunately, he was not paralyzed.[1]

None of us knows what will happen when we leave the house in the morning.

As Nancy Pelosi once said: "Every family in America is just a phone call away, one diagnosis, one accident away, from needing the benefits of stem cell research."[2] Wise words indeed!

For my vote, I suggest we have cures ready and waiting, before disaster strikes.

It is like the story of the General, whose army was defeated in a crucial battle. The survivors were just about to scatter in all directions.

But the General convinced them to stick together, to trust him one more time. They were hungry, cold and shivering, out of supplies — but they still trusted him — at least for one more time. He told them to head in one direction, and they did.

And when at last they came to a particular hill — and an apparently abandoned castle — they found it was packed with supplies, food and warm clothing, weapons and more. The General had planned ahead: whereby his soldiers lived...

As of this writing, the California stem cell program is running out of money.

---

[1] https://www.golfdigest.com/story/how-a-trio-of-tour-pros-learne
[2] https://pelosi.house.gov/news/press-releases/pelosi-with-great-potential-of-embryonic-stem-cell-research-science-has-power-to

Unless our state renews the program, CIRM will die; the California Institute for Regenerative Medicine will cease to exist.

Would Bob Klein lead the effort to renew?

I remember where I was, when I first heard him speak to that. It was in his office, and I had just handed him a sheet of paper with some good news: CIRM had approved a $6 million grant to support more rapid bone healing.

I was just beginning to be aware of age. Age...when the bones become thin, even to the point of fragility. And if you fell? a younger person might bounce right back up. For an elder, that same fall could be devastating. Breaking the hip circle could be fatal, or lead to being institutionalized. Instead, CIRM might fund ways to strengthen the bones, and/or accelerate healing.

With my gift of stating the obvious, I said:

"That could be something wonderful."

Bob nodded — and then seemed to go deep in thought. I waited. The pause lengthened.

When he spoke, it was softly, as if I had gone away, and he was alone in the room:

"The scientists are worried about their funding... but I will get it for them."

To continue CIRM, Bob intended to ask California voters for $5.5 billion. That would give life to the program, lasting perhaps to the end of my life, and beyond. It will be no easy task, persuading the voters to back stem cell research again; it would be the greatest stem cell battle of all time.

Remember Lee Trevino, struck by lightning? Some nerve conditions (like stroke) do damage like a lightning bolt striking from within. What if it could be prevented?

The hoped-for program sets aside $1.5 billion for "Diseases and Conditions of the Brain and Central Nervous System"...

Could CIRM deliver on such a huge promise? Follow the money. What had been done already?

How much has CIRM actually spent on three conditions related to the brain and central nervous system? See below, from the public record, www.cirm.ca.gov:

Parkinson's: fifty-seven million: ($57,612,450.00)

Stroke: sixty-two million: ($62,221,206.00)

autism: fourteen million: ($14,620,319.00)

And if we DON'T fix these giant problems, how expensive is that, in annual costs?

Parkinson's disease cost $52 *billion in one year*(Michael J. Fox Foundation)

Stroke: $46 billion (CDC figures)

Autism: more than <u>$268 billion a year</u>.[3]

The cost to America of these conditions is measured in the **billions** (with a "b"); but cure research by CIRM is only (only!) in the **millions** of dollars.

It is more expensive (by far) to live with a disease than to fight it.

"What can't be cured must be endured", the old saying goes — I vote for cure. And if we can prevent it, that is the best of all.

Or, we could do nothing — and hope the lightning does not strike too close.

---

[3] https://pubmed.ncbi.nlm.nih.gov/26183723/

# 16 The Man Who Fell from the Sky

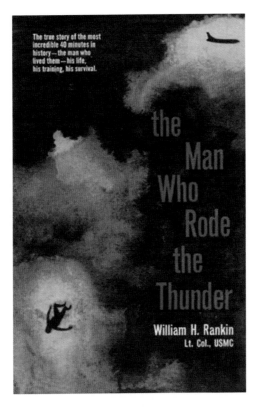

His jet on fire, Lt. Col. William Rankin fell twice the distance of Mt. Everest — into a lightning storm. (*The Man Who Rode the Thunder*, Rankin, Amazon.com)

I love the possibility of miracles, and hope to experience one before I die. But first, let us agree on definitions. To me, a miracle is something which seems impossible, but which happens anyway.

I'll tell you what I want in a minute: but first, a physical miracle, by almost anyone's standard.[1]

On July 26, 1959, Lt. Col. William H. Rankin was piloting a jet (an FBU 9) 47,000 feet above Cherry Point, North Carolina, when he heard a thump and a rumble from within his aircraft. Worse, in bright orange letters, a warning light flashed on the dashboard: **FIRE**.

Rankin was flying incredibly high, at the edge of space. Far below him was a gigantic black cloud, with lightning bolts bursting out of it. No problem, he would just fly over it.

Except, he suddenly realized, he had no power.

He pulled the auxiliary power lever — it came *out of the wall* and fell on the floor.

He radioed his partner, flying nearby, saying: "I might have to eject."

"If you go, let me know," said the other.

Rankin was almost ten miles up, above the world. The temperature inside his cabin was a comfortable 70 degrees. Outside? Seventy degrees — *below zero...*

He pushed the button marked EJECT. Instantly, he was outside. The jet diminished in the distance and was gone.

Burning cold. Where skin was exposed, instant frostbite.

One glove yanked off and disappeared. He used the other hand to keep his mouthpiece shoved in, so he could breathe, from the airtank on his back.

His belly expanded, agonizingly, as though he was pregnant. (This was due to the difference in air pressure, inside his body being much greater than the airless atmosphere.

An invisible giant seemed to play catch with him, tossing the pilot high, then dropping him violently, to be caught and tossed and flung again, changing direction, hurtling hundreds of feet in seconds.

His parachute deployed, but it brought him no control. The forces around him were too great.

And below him was that giant black cloud.

He fell into the electrical storm. Huge bolts of lightning struck beside him, colors and shapes he had never imagined.

Rain, so thick he feared he might literally drown.

---

[1] "The Man Who Rode the Thunder" by William H. Rankin, is included in the book WILD BLUE by David Fisher, available at Amazon.com.

Then he was flung higher in the storm. The temperature dropped violently, freezing the rain into hail, which struck like bullets, welting his body, dark bruises all over.

Wind threw him upwards, into the chute, which covered him like a shroud — then dropped him — entangled, disentangled — all in instant.

He fell and fell, faster and faster, below the storm — he felt the tug of gravity — saw a patch of green below...

A forest. He might be impaled on a sharp tree branch.

He slammed against a trunk, broke off branches, hurtled to — *the ground.*

He somehow managed to stand up, and unclip his parachute. He staggered to a road, stuck his thumb out, attempting to hitch-hike — but passersby were few, and he looked strange.

Nobody would stop and pick him up — until a little boy in the backseat of his father's car said, "Daddy, that man is a pilot, he is in trouble, stop and help him!"

Lt. Col Rankin later gave the boy the helmet he was wearing that day.

He had fallen a distance roughly twice as high as Mount Everest.

His jet? It crashed into a farmer's field, exploding in a ball of flame...

Bill Rankin lived to the ripe old age of 86, respected and revered by all.

I want a miracle like that: something incredible, seemingly impossible, but just barely doable.

I want my paralyzed son, Roman Reed, to walk again. It has never been done before, not in the history of the world. No one has ever come back from quadriplegia, paralysis in both arms and legs. But that's what I want. That is the miracle toward which I work, every day, and have, for 26 years.

And when one day it happens, and Roman is standing between the parallel bars, sweating from exercise, we will pause to honor the memory of William H. Rankin: and lift a glass of apple cider to the man who fell from the sky.

# 17 Tunnels

Yasmin Wengro was our family doctor since the birth of our children. It gave her much pain to say that Gloria had cancer of the pancreas. (health.usnews.com)

Gloria and I were lying in the dark, with our eyes open.

"I have a stomachache," said Gloria, " But the pain is also in my back. I think it's cancer."

"No, no, no — " I said, "Lots of other things it could be — like kidney stones."

"Those hurt when you pass them," said Gloria.

"You smash them with sound waves first, they get really small!"

"How do they do that?"

"I have no idea," I said, "Besides, maybe it's nothing!"

"No," said Gloria. "It is not nothing."

The X-rays did show kidney stones, one of them "big as half your thumb", said the doctor on the phone.

This was Dr. Wengrow, a cheerful African-American woman who had been our family doctor since our son Roman was born. We trusted Dr. Wengrow.

I smiled at Gloria, gave her a thumbs-up, like, see? Just kidney stones! You're fine.

"But there's something else," said the doctor, and a chill went up my spine, "Out of an abundance of caution, I want to check on it. It is probably nothing, but still I want to be sure."

There was a spot on the middle of Gloria's pancreas.

A black spot. I thought about TREASURE ISLAND, by Robert Louis Stevenson. When the pirates wanted to get rid of their leader, they tore a page from the Bible and wrote a black spot on it, from a burnt piece of wood.

The black spot. The death sentence.

At Kaiser Medical Center, Gloria lay on her back in the MRI tunnel, one of those noisy BONGBONGBONGBONG things that takes detailed pictures of your insides. I stood close to her, one hand touching her ankle, as we waited for the machine to turn on.

"Remember London?" I asked her.

"Tell me," she said, relaxing a little. She liked my little stories.

We had been supposed to go with friends. But at the last minute, when Gloria and I were actually on the plane, the phone call came — they were not coming. And they were the ones who knew London! We were on our own.

I had arranged for Roman's medical care, two nurses who would see him every day. We could not afford the trip, really, but how much time did we have left, both of us being around 75, at the ¾ century mark

London had always been on Gloria's "bucket list" — to see the Tower of London and the crown jewels.

For me, England was a chance to experience "THE WIND IN THE WILLOWS", by Kenneth Grahame. At the "River and Rowing Museum", on the River Thames, they had a special exhibit on my favorite book of all time. It was a tunnel, supposedly populated with oversized statues of the animals. I had heard about the museum years ago, and always feared it would be taken down before I saw it. And now, here we were, just a short train ride away.

Only, it was not going to happen. To get there, I would have to travel through and beyond London — by myself. Gloria's knees were not up to it. I had asked about disability conditions, and the front desk folks said there was a lot of stairs and way too much climbing.

But that was not the reason I changed my mind. I was just scared to do it by myself. If Gloria was with me, everything would work out. She was the practical one, she would manage things. That was how it was supposed to be. But if she could not come along? I would just get lost!

I told her I was not going, that was it, I had changed my mind.

"Oh, no!" said Gloria, "You have been talking about this museum for years. This is the only chance you will ever have — you are going."

"No," I said, putting my foot down, "I'm not."

The train rattled and shook as it raced through the English countryside.

When we came to my stop, there was no one else in the train car. I went to the door — but *it would not open*. I <u>knew</u> something awful would happen, if Gloria was not here — I was rehearsing what I would say to her as I pushed all the buttons over and over — PSHSHSHSH — it opened.

A mile's easy walk through ever more amazing green countryside, a river with willow trees, branches gently bending to the ripples, just like in the story — and there it was: the actual museum, with a big sign, "WIND IN THE WILLOWS".

There was a lot of history on the walls inside, how the book was written to amuse the author's son, who had a chronic disease. The boy's personality had inspired the boisterous Mr. Toad. I read the signboards, of course, every word. There was also a Winnie the Pooh exhibit, which I ignored — that was for children!

I stood outside the tunnel, practically holding my breath. Before me was a friendly dark, pregnant with possibilities. But after all these years, had I built it up too big in my mind, so it could not live up to expectations? I entered.

On both sides were big windows, and behind everyone a lighted scene, with statues and painted backgrounds from this most wonderful of books.

I went slowly: lingering, savoring, breathing it in.

There was Ratty, the literate river rat, who wrote poems, and Mole, that most loyal of friends, such adventures they shared: from tipping the boat over on their first picnic to getting lost in the Wild Wood, and finding the doormat (!) in front of Badger's snow-covered home, whereby they were saved.

And Mr. Toad! There he was, going to jail for "borrowing" a motor car, and in his absence the weasels stole his ancestral home — and when Toad escaped, the brave four animals had to fight to get it back — wonderful...

I went all the way through the tunnel, slowly, very slowly, studying every window. And then — I did it again!

On the train ride back, a smiling old lady sat down across from me.

"I know where you have been," she said, pointing to my WIND IN THE WILLOWS shopping bag, bulging with every "Willows" thing the shop had had to offer, including a banner which hangs above my desk to this day.

"*I helped to paint the tunnel you went through,*" she said.

It was a little blessing from the Wind in the Willows...

"I can top that," said Gloria when I had made it back to our hotel room without further adventure.

She had gone for a swim in the hotel pool.

Which was all very well, four feet deep, no fears of drowning — but then she realized at some point she might want to get out — and the stairs were too high on the wall. Getting in was no problem, gravity helped — but to climb out? Her arthritic knees were not up to the challenge.

There was a young couple in the pool, and Gloria forbade them to leave.

"But — we have to go," they said, somewhat puzzled by her commands.

"No, you have to go get someone," said Gloria, "Or I will be stuck here forever".

When the well-dressed staff appeared, she explained the situation, and they got a mechanical hoist...

The crown jewels were a disappointment. To me they looked like gawdy plastic jewelry. I have never understood the lure of shiny rocks. What makes a diamond more valuable than a river stone? Just opinion, really.

Gloria shook her head, saying: "Think what all that wealth could do for the hungry and the homeless".

We sat in Shakespeare's schoolhouse, where Earth's greatest writer had studied. I went to the Sherlock Holmes museum, with its villainous

Moriarty statue, and we rode on the Eye of London, a slow, gigantic Ferris wheel with a magnificent view of this great city…

And the Tower of London somehow glowed blue-green when we saw it…

"Ahhhh…" said Gloria, and that was her favorite moment.

BONGBONGBONGBONG — "Here we go," said the voice in the hospital, and Gloria slowly disappeared into the MRI tunnel.

I maintained my grip on her ankle, holding it the entire time.

"I liked that," she said afterward, "That you were holding me. If something went wrong, you would pull me out, not let anything bad happen to me."

She paused.

"When I am dying," said Gloria, "That's what I want. My grandmother died in a movie house, by herself, all alone — I don't want that," she said.

"When I cross over," said Gloria, "I want you holding my hand."

"I promise," I said.

# 18 Grampa and the Lion

Not long ago, in semi-wild Africa, an old man lived with his grandson, whose parents had died.

They enjoyed each other's company, and were content, until one day when a lion entered their hut. It was a fully grown male, but past his prime, and parts of his mane had fallen out. He was perhaps no longer capable of catching his normal prey, which made him dangerous.

Grampa leaped on the back of the lion, wrapped his skinny arms around the animal's throat, and shouted instructions at the boy, who ran next door.

"May I borrow your gun?" he asked.

Fortunately the neighbor did not ask a lot of questions, just handed him the gun, telling him it was loaded, and clicking off the safety.

The boy ran home, placed the barrel of the gun on the roaring lion, where he hoped the heart would be, and pulled the trigger.

By freak luck and courage, Grampa did not die. They skinned the lion, and the villagers helped them eat it.

Question: what would have happened, if Grampa had not jumped?

That was roughly the question California had before it: to renew (or not) the stem cell program's funding.

We could decide to do nothing, and hope the diseases would not come our way — or we could continue the fight.

Remember when our program began? One scientist (whose name will not be mentioned) predicted we'd be lucky to get even *one* therapy to human trials.

I understand that prediction. Our first therapy (for spinal cord injury) took almost 20 years to reach human trials.

Many (perhaps most) therapies get eliminated early. They do not survive the basic experiments, the test tubes and petri dishes, and that is the cheap stuff.

The expensive part is the "Valley of Death" middle part, the seemingly endless rat and mouse tests, when there is almost no funding available. Nothing is for sure, it is all visionary, forward-looking, into an unknowable future. It might be wonderful; it might be nothing. All that is sure is that it will be expensive.

The last part, clinical trials, seldom happens until there is almost a guarantee of good results. Only then are investors likely to jump on, to get a piece of the action.

But that in-between time, years and years of tests — that is when it's hard to get funding. If a therapy does get past that, to clinical trials, for human volunteers? That is a very big deal indeed.

To **not** continue such science, after we have gone through all the hard years, survived the Valley of Death, all the way close to clinical trials — and to never know what great things might happen?

That would be like stepping off the lion's back — before the kid came back with the gun.

# 19 Diagnosis and Decisions

Irv Weissman's discovery of the cancer stem cell may be the answer to cancer. (Stanford. edu)

There were more tests for Gloria: including something called an endoscopy.

The endoscope (a tube with tiny scissors on one end) was pushed down Gloria's throat (she was asleep) all the way to her pancreas. It snipped off a sample of the dark spot, and brought it back for analysis.

We had to wait ten days for the results. I took that as good news, that there was nothing to worry about, "No news is good news!" Gloria felt the opposite.

Then Dr. Wengrow's nurse called for us to come to her office, whereby we knew it was bad. Good news can come in a phone call, or an email, just "You're fine, have a good day." — but if we had to come in?

We held hands in the doctor's office, and waited for her to come in.

"I am so sorry," said Dr. Wengrow, as soon as she came in the door, "It is cancer."

Cancer of the pancreas is deadly. Just finding trouble is hard, the pancreas being hidden behind the stomach. There may be no warning symptoms, until too late.

"How long do I have?" asked Gloria.

"We can't know for sure," said Dr. Wengrow, "With chemo and successful surgery, some people can go on for years."

"And if we do nothing?" said Gloria.

"Two to four months," said the doctor, trying to blink back the tears.

I reached out to the strongest people I knew, giants in their field: like Irv Weissman, who discovered the cancer stem cell; Catriona Jamieson, leukemia expert; Dan Kaufman, of natural cancer killer fame, others — and each one asked for more information, which I provided.

I spoke with Bob, of course. Not only is he the man who began Prop 71, the California stem cell program but his wife, Danielle, had survived breast cancer. In the field of cancer cure, Bob knows everybody who is anybody.

He asked me first who had I called, and what did they say? I told him. Then he said:

"I'm sorry, Don."

In that moment it became crushingly real, like a great weight falling onto my shoulders. If there had been any hope at all, Bob would have named the scientist who might best help, and have made a phone call to introduce us.

He volunteered to pay for acupuncture for Gloria, to help with the pain — and for me, to ease the stress. This was no small gift; acupuncture costs $85 a visit.

And then he said: "Do your work at home now. Stay with her."

This was before Governor Newsom sent the whole state home, because of COVID, this was just an act of kindness on his part.

He hugged me, the way I had hugged him when his son Jordan died.

I gathered my stuff and went home.

Of course, I kept trying.

I repeatedly visited the **clinicaltrials.gov** website, typing "pancreatic cancer" into the search box, studying the tests that had "recruiting" beside them.

But there were always conditions, endless restrictions on who they could accept for the tests — and most of them disqualified Gloria.

There were three main strikes against her.

Her age? 75.

Her health? Overweight.

And the state of her cancer? Too advanced. Her "cancer number" was supposed to be 30 or 40. Hers was over *two thousand*: 2,167.

One doctor said: (Gloria's) "cancer has mutations (KRAS G12V and p53). Unfortunately, there are no targeted drugs for these mutations…"

Hans Keirstead said he could put in stem cells which might help, but only if there was a "resection" surgery; in other words, they would have to cut Gloria open. But that could not be done, our doctors said. The cancer on the pancreas was too "involved" with several arteries, and the surgery was probably not survivable.

Another cancer specialist said Gloria was now too weak for "aggressive chemo and surgery"; she would have to be strong enough to be on her feet 50–60% of the day, which she was not. This particular doctor spoke into her phone, recording her diagnosis.

"She is visibly weak, has constant pain and nausea, and spends most of her day in a reclining couch," she said. I sent mental suggestions to Gloria, "sit up straight, look strong — " but she could not.

In good conscience, the specialist said, she could not make Gloria worse, increasing her pain. Surgery, in her opinion, would involve parts of the surrounding liver, bowel and spleen, as well as the pancreas. Recovery would be rigorous, lengthy, and painful.

Her recommendation? Hospice, a form of assisted dying.

However, the Department's head surgeon said he *would* authorize an operation for Gloria, if she wanted it. But he personally did not recommend it. His own mother had died of cancer, though he had been able to keep her alive for three years after diagnosis. He would not have done it again, he said. The pain had been too great, and the quality of her life was not good.

For a while it looked like we might get Gloria into a clinical trial in San Diego. But after studying our information, the scientist said: "…For this particular trial…the patients' cancer has to be stable over a period of at least 4 months on chemo — so she would not be a candidate…" Gloria was on a mild form of chemo, a drug called gemcitabine.

Gloria refused anything that might increase her suffering, or prolong it. We began discussing things nobody wants to talk about.

"When it is time," said Gloria, "I want to go to the hospital".

"That does not make sense. Your home is here," I said.

Half a century ago, when Gloria and I were first courting, there had been a moment when I felt I was being watched, and tested.

Gloria had a nerve condition in her hands. Every so often they would suddenly perspire. I loved to hold hands with her, but for a long time, she shied away. Even so, it was natural for me, and one day when our fingers were intertwined, I felt the sweat pop out from her palms. I had the feeling that she was watching me, and I should be careful what I said next. So I said nothing, which turned out to be the correct response. When this happened with other prospective boyfriends, she told me later, they would say something stupid, like "I must be hot because I turn you on". But I kept my mouth shut, whereby she knew we had a chance as a couple.

Why would she want to die in a hospital?

It took a while before she would say.

Generally, Gloria was the boss on the practical side of our life. She did a great job, why wouldn't she be in charge? On major decisions, we worked together, figured things out. She knew, for instance, that I was going to write in the mornings, every morning, and she did not mess with that. But most things were up to her. I asked what she wanted; that's what we did.

"It is going to be a lot of trouble — too much trouble," she said. "We'll just let the professionals handle it; that's why we have insurance." She looked at me like the discussion was over, decision made.

To send her away, to die among strangers, instead of we who loved her — because it was... trouble?

"That is not an issue," I said, taking her hand.

Whatever came, we would face it together.

# 20 The Possibility of Cure

Author playing Tai Chi sword style by Niagara Falls.

I have, first of all, the greatest job in the world. Since 2003 I have worked for Bob Klein: beginning as a volunteer, presently as vice President of Public Policy, for Americans for Cures Foundation.

Every morning I get up around 2:30, and write for a couple hours about stem cell research. I write seven days a week and do not recognize holidays. People who work miserable jobs should absolutely have holidays — but I am happy with my job, so why would I want days off?

Also, every Tuesday, Thursday and Friday I drive across San Francisco Bay, pull into a Palo Alto parking lot, lug a briefcase up three flights of stairs, and work in Bob Klein's office from 7:30–1:30. At home after that, I will read my notes for an hour, getting ready for the next day.

At first I used one of the Klein Financial real estate offices, on a borrowed computer. Bob told me once he was going to build an advocate's office for me. I thought he was just kidding, but Bob does not joke a whole lot. One day I came in, and a stranger with a sledgehammer was pounding holes in the wall. The next day, there it was: an office, small and cozy, book shelves to the ceiling, and a landline phone. It was perfect.

Until COVID struck, I attended nearly every meeting of the CIRM board, sometimes two or three a week.

In short, I was doing exactly what I wanted to do with my life, working with people who were (much) smarter than me, trying to find ways to encourage cures.

Now I was working from home.

But — are cures possible?

Years ago, I had a headache: a particularly savage one, right behind my eyes, as if some small fierce animal was inside, trying to claw its way out.

It was right before Tai Chi practice, which I was heavily into at the time, trying never to miss a practice. It was at Fremont's Lake Elizabeth. Our leader was International Champion Mai Chen. She had a purple silk outfit, and she moved like flame.

I was early, sitting at a concrete picnic bench beside our practice ground, when one of my Chinese friends came up. Her name was Chia Yu, and her English was about as good as my Mandarin.

I went to a Chinese movie once, and in the whole two hours I only understood one sentence, which was: "Wo bu zhir dao", which means: "I do not understand."

But Chia Yu knew I was trying to learn Mandarin, and would talk at me sometimes. I seldom understood. But I would nod my head and smile, and repeat the sounds back to her, and it was a happy time.

Today, a couple words occurred to me. I said, "Woh to-tung". This meant, I hoped, "I have a headache". Or, perhaps, "I am headache".

Chia Yu looked at me, opened her purse, and whipped out a little bottle of green pills which I hoped were the Chinese equivalent of Advil.

I reached for the bottle.

She shook her head, pulled the bottle out of my reach, put it away.

Then, with her right hand, she gripped her own left wrist. She said something, repeated it, slowly, as for a not-too-quick child. I had no clue. I tried saying it back to her, and she frowned, shook her head. Then she tapped her ankle.

Then another friend came along, who spoke Mandarin, Cantonese and English very well.

Second Friend smiled.

"She says, your blood pressure is wrong; she suggests: take off your shoe."

So — I took off my shoe.

Chia Yu took my stocking'ed foot, made a pincer out of her thumb and forefinger — and sharply squeezed my big toe, a hard scraping motion.

The headache went away, instantly, like turning off a light. I was still slightly dizzy and light-headed, but the pain was gone. It just wasn't there anymore, and it did not come back.

"Acupressure", someone said, putting a name to the kindness of Chia Yu.

Why does this matter?

In a small way, I had just experienced cure.

Cure does exist; it is an actual possibility. People can get well.

Let that be written as an article of my faith.

But could I save my wife?

When Gloria and I went to Bob's acupuncturist, and had the painless little needles inserted, it did help: with her pain, and my stress. It was a beautiful gift.

But the cancer did not go away.

# 21 Gloria's Last Request

At my life's low point, Dr. Anne Marie Duliege gave me an unexpected gift. (blog.cirm. ca.gov)

A stem cell friend, Dr. Anne-Marie Duliege, inquired about Gloria's and my situation, and offered an amazing act of kindness.

She was worried, she said, that if Gloria passed, that I would go through the rest of my life feeling there was something undone, that I could have saved Gloria, if only I had done X, Y, or Z…

She had two pancreatic cancer expert friends, she said, one in California, one on the East Coast. Could she contact them, for their opinions?

Naturally, I jumped at the chance.

She sent each friend the following letter:

"Dear ____

"One of my closest friends is Don Reed; (we work together on promoting stem cell research); his wife Gloria was diagnosed with locally advanced unresectable (non-operable) Pancreatic Cancer.

"Gloria is 75 years old and not in very good health unfortunately (overweight, etc.). Given her poor prognosis and what's at stake, I suggested to help him by getting a 2nd medical as well as surgical opinion. He is totally on board with that, and is grateful for my initiative.

"If you are OK with giving me your opinion, I will send you additional information, as needed.

"I really have only 1 question: Do you think that a surgical intervention is out of the question at this point? This is the position of Don's health care provider, as Gloria's cancer has splenic artery involvement, encroachment on the celiac axis and splenic vein thrombosis. And this position makes sense, of course.

"However, Don would be ready to consider (and of course discuss with his wife) any approach that is slightly more aggressive (within reason) than a conservative approach.

"Would this be agreeable to you?.

"Best regards, Anne-Marie Duliege"

I obtained and sent information to the two scientists/doctors.

The answer came back. The emphasis is theirs.

"Dear Anne-Marie:

"...The options are clear here. She definitely has INOPERABLE pancreatic cancer and should NOT have surgery. With the CA19-9 (cancer number) in the thousands, (2,167) she has incurable disease.

"Only option is chemotherapy — (but) she has a very poor functional status which predicts that she is very UNLIKELY to respond to chemo. Were she my sister, I would recommend hospice. As she is already bedridden and miserable, we really have nothing to offer except comfort.

"I am sure this is extremely difficult for Don who is into stem cells and curing the impossible. But there is no other option."

Their words were kind, and intelligent, like Dr. Duliege herself. Her gentle purpose, that I would know there was nothing more I could do — that was accomplished. But what I really wanted was hope, and there was none to offer.

The research had not advanced enough.

Several months before we knew about the cancer, Gloria had fallen.

She had been on her way to hang laundry outside, in the early morning, when it was still dark. She tripped, and broke her collarbone terribly, and never fully recovered.

So now, when she sat down a lot, it seemed at first that her shoulder was just giving trouble. Maybe she was just resting, and would soon be her usual mischievous self again.

She even pulled a little trick on me, connected to the injury.

I knew she loved to have me go to church with her, (which I rarely did, not being of her faith, nor of any organized religion) and so I told her, as long as her shoulder gave her pain, I would take her to Church.

Every Sunday as we walked up the aisle toward her favorite bench, I would gently support her stronger arm. Once sitting down, I would stabilize the other side, the wounded limb, partially lifting it throughout the whole service. It was exhausting, but seemed to give her some relief.

I did not find out until later that her shoulder had healed to a nearly pain-free state. She just liked having me beside her in church, and never got around to telling me.

Energy was always her defining feature. But now in church she could barely stand up for the prayers, and would sit down if the Priest went on too long.

She who was so strongly anti-drug was now forced by pain to take morphine, as well as stuff with names like agravan and lorazepam for the nausea caused by the heavy-duty drug. Pain-killers have side effects, like constipation, and we took endless medications to deal with that.

It was a time of last things. Some were fun.

I will never forget our last game of cards. We used to play every night, and she was amazing at it. She could not only keep track of her own cards, and the ones she had discarded, but also the ones that I took from the face-up pile.

She could tell me mentally what card to put down, according to her need. I would feel a warm tugging at my brain, like I really wanted to put down the three of clubs, and I would do it and she would use it. Sometimes I caught her at it, the mental equivalent of a kid with her hand in the cookie jar, and I would say, "No, I am not going to put down the three of clubs!" And she would laugh and do it again, after I relaxed and my guard was down again.

We played for 25 cents a game, and the loser (guess who) would put the quarter in the vase in the living room. When there was enough, she and her sister Brenda would go to Reno and play Pao Gao.

But it was getting harder for her to play, almost impossible to get up from the couch and be supported on the short walk to the kitchen table.

On that last game, somehow I kept getting the most wonderful cards — it was almost impossible for me to lose. But she loved to win so much! So I played even worse than my usual, stupider and stupider until I somehow found a way to lose — and what a joy it was to hear her laugh once more. (Desiree thinks she was trying to let me win, but I don't know.)

Our biggest problem over the years (one of the few things we quarreled over) was time spent together. She always wanted more of my time, and I was always busy on one writing project or another.

But now was our last time to be together, and I did not want to miss any of it.

I got up in the morning at my usual 1:30 AM (I went to sleep at 7 PM, so it worked out the same as most folks, just arranged differently.) I would write for a couple hours, and then (when she cleared her throat a certain way) come back to bed with Gloria, and we would talk till daylight.

As long as she could, I would walk her downstairs, carefully escorting her all the way, past the big doll house where our grandaughter Katherine used to play, past the brass diver's helmet Roman gave me, below the glass-etched painting of me "walking" the white shark, from the old days as a diver at Marine World. Gloria made that picture for me, scratching it onto a window pane. She had only done the one picture, and the instructor said it was too difficult for her, but she did it anyway. And last was our family photo wall: Gloria and me on our wedding day, Roman looking huge and strong, and Desiree and Josh, the day their hands were joined in wedlock.

Step by careful step we went, and then across the floor, to install her in her corner of the couch, beside the fireplace. There she would stay, ensconced like a queen, barely moving at all, until night came.

There came a day when Gloria said, "Enough, no more!" She flatly forbade me to do anything else. She had been on a mild form of chemo; she stopped its use.

Naturally, I did not agree. But the final decision was hers; it had to be, and I respected it.

She officially requested hospice, a program to ease a person's last hours.

She was not afraid of dying, Gloria said, she was going to be with God. I said all well and good, but what about me? We'll only be parted for a little while, she said.

And then I knew the one thing that would give her joy.

I signed up for instruction in the Catholic faith. I told the Priest I believed in embryonic stem cell research and always would, no matter what they made me say in the signup ceremony. He said that was all right, the church was wrong about the gay rights issue too.

Things had to move fast so Gloria would be there, to see me inducted as a Catholic. She could barely sit up, but she was going to be there.

She told the kids, don't say anything to Dad, if he feels pushed, he might not do it. She told Desiree's husband Josh she was glad I was converting, because if I was a Protestant, as I was raised, she might not be able to locate me in Heaven.

The priest told me to pick a saint's name, to put beside my own. That obviously had to be Joan of Arc, my inspiration in life. He asked was I sure? And I said Yes.

And that was why, when I was later introduced to the congregation, it was as "Don Joan of Arc Reed".

Gloria smiled, from where she sat, though she barely had the strength to hold her head up.

We were together all the time now. I talked and talked and almost never stopped, except sometimes she would put her hand over my mouth, which I took as a hint.

At first, I had "cooked" in the most generous definition of that word. Gloria sat at the kitchen table and would tell me what spices to use.

But as her condition worsened, she lost interest in eating.

Then she could no longer navigate the stairs. The hospice folks brought her a hospital bed, in which she stayed.

When she had to use the restroom, I would sit her up, put my arms around her in dancing position, and waltz her backwards till we got there.

I read aloud to her, first just one paragraph, then (at her request) the entire book of "WIND IN THE WILLOWS". "Read it all," she said, and I could feel her satisfaction at sharing something which meant so much to me and for which she had sent me to the museum. So I did, and that was a time of happiness.

It felt as if I could just keep talking, she would not leave.

Memories flooded back.

Like when the dive department of Marine World built a playhouse for Desiree and baby Roman. It had been a lion's "lie down bin" originally, now painted into a beautiful little white house with green trim, and an address on it, 40480½.

Desiree and Roman could not quite believe what was happening. Like every great Mom, Gloria had just always been there, part of their lives, giving them mountains of attention and oceans of affection, all the love and guidance she could give. And now she would not be there?

Desiree kept calling, from her workplace in Nevada, where she was Athletic Director for the University of Nevada at Las Vegas, asking: was I sure, was this real, Mom could be dramatic sometimes, was she really... dying?

"Remember when baby Desiree first swam?"

In bed, Gloria nodded. She did not speak; her eyes stayed shut.

It was a program where they just put the baby in the pool, and let her swim. "naturally". I understood what was going to happen, but apparently Gloria did not. They dropped Desiree in, and I could see her underwater just starting to "climb", making the swimming motions — Gloria freaked out. She leaped in and grabbed her daughter — though she herself could barely swim. I had to jump in as well, and rescue both of them. It is hard to swim with your shoes on, BTW.

When we got out, Gloria handed Desiree to me, and went over to the instructor and *socked* her in the chest.

In the bed now, Gloria made a light breathing noise, like laughing.

Such joy she gave me, and what a powerful defense when I most needed her.

It was Gloria who got me my diver's job at Marine World. She had been 7½ months pregnant at the time, and I was in Louisiana looking for oil field dive work, and so she did the job interview for me. Halfway through, she started to cry, and said, "He's a good man, you just have to hire him!" And the head diver said, "Lady, lady, I will give him the job, just don't have your baby here!"

"Remember when our kids and I graduated from college — Desiree from UCLA — she used to call it Yoosalay — and you were so excited when she went up for her diploma you could not even see her!

"And when Roman graduated from Cal Berkeley, and they said his name, ROMAN REED, it was like written in letters of gold across the sky.

Gloria had even made me take college courses, one at a time. I never knew why, until finally she said: "Now is when you become a teacher."

I said I had no interest in becoming a teacher, to which she responded, "Let me put it another way. Now is when you become a teacher — or I divorce you!"

She said later she was only joking, but I wasn't about to test her.

"And when I graduated from college with a Teaching Credential..."

During the ceremony we were supposed to just take the certificate and move on, keep the line moving, but I stopped the procession and said:

"This would not have happened without the endless encouragement of my wife, Gloria Jean."

She also told me when it was time to retire from teaching, and work full-time on stem cell advocacy.

And now?

In one of her last coherent sentences, she said:

"They have to pass this stem cell stuff; nobody should suffer like this."

Which I took as a last request.

# 22 The Pope and the Whale

Tahiti et ses Iles
*Baleines à bosse*

And then the baby whale poked its head out from under its mother; "baleines a'bosse" (artist unknown; if you know him, please ask him to contact me.)

It was Easter Sunday, years ago. Gloria was alive and healthy, and she and I were standing in St. Peter's Square, in front of the Vatican, at the very heart of Catholicism: Gloria's faith.

In those pre-COVID days, thousands were jammed together, to see the Pope.

We had gotten there hours early, wanting to get as close as possible to Pope John Paul the Second. This was before he had been nominated for Sainthood, but everybody knew he was someone special. He had

Dr. Maria Millan: becoming the President of CIRM. (blog.cirm.ca.gov)

that glow of connection with people, as if he was someone you had always known, and who was on your side.

I disagreed with him on the stem cell issue, of course. Those differences were probably irreconcilable.

To me, embryonic stem cells (ESC) were something shining and wonderful, which everyone should support. But the Catholic church? Not at the top level. The members, however, were generally in favor. Perhaps they had been propagandized so much by the church, that they became interested, and studied the issue, and made up their own minds.

"A Time poll conducted by the SBRI research group in June of 2008 found that 73% support embryonic stem cell research..." — that is a lot of Catholics![1]

In another poll, a large majority of Catholic voters (72%) supported "allowing scientists to use stem cells obtained from very early human embryos to find cures for serious diseases such as Alzheimer's, diabetes and Parkinson's."[2]

The Catholic Church has long been considered an enemy of science.

---

[1] https://www.theatlantic.com/politics/archive/2009/03/stem-cell-polling-support-has-steadily-grown/1336/

[2] https://www.catholicsforchoice.org/press-releases/catholics-support-stem-cell-research/

Michaelangelo, who painted the Sistine Chapel, had to dig up dead bodies to learn anatomy, a crime in those days. Leonardo Da Vinci actually had a professional grave digger on his staff, to provide him with corpses to study. If they had been caught — the genius artists might well have been hung!

I also objected to the lesser status of women in the church, forever denied a leadership role as priest or pope; it was said that once, long ago, there had been a woman Pope, but this was not certain.

Gloria shocked me in Rome, the night we went to the Colosseum, and were jammed into a crowd.

Gloria punched a woman in the face — and the woman had a baby!

Or so I thought, until I saw two hairy arms come out from under Gloria's blouse. The baby was a doll in a sling, so the man's hands were free to pick pockets. He had tried to get Gloria's chest-pouch, but my wife was not having it. She punched, and he took off, clutching his wig with one hand, the plastic doll with the other...

The Vatican itself was astonishing. We loved the paintings on the walls, some parts Michaelangelo's work, and some parts not; the Sistine Chapel was amazing.

Suddenly, as we waited outside, the Pope was there, in his little white "Popemobile". That is not rudeness, by the way, but its official name. He had the top down, waving at everyone and coming closer and closer — I stuck out my hand, leaning over the rope, far as I could reach, standing on one foot, just about to fall — and the Pope saw what I was doing, and stretched out his hand too.

Our fingers touched — he gave the famous smile and moved on.

I turned to Gloria and said, "Skin cells, skin cells!" and touched her hand with mine, which had just touched the Pope's...

It was, Gloria said later, the highlight of her religious life. She did not know it, but she would one day give me mine.

For me, the ocean is a holy place.

We were in the Bahamas. Gloria had been saving for a vacation. We did not have to pay rent, because we stayed at a friend's house. It was very much a country place, overrun with chickens (don't leave the door open!) beside some of the most beautiful ocean imaginable.

One morning I swam with lemon sharks, big and beautiful, skin pale yellow, eyes of shimmering gold. They approached me, and I could feel their gaze evaluating me — was I food, furniture, or something that could hurt them? That is about all a shark has time for. Then the golden eyes flicked ahead, the sharks moved on.

But that was not the moment.

Against my will, we went on a whale-watching trip. I view such trips as generally a colossal waste of time. A big success is a fin tip cutting the surface, and everybody ooos and ahhs, and some describe it as a religious experience.

But this time we saw — nothing. Just empty water. See? I told Gloria.

When we got back to the dock, Gloria practically assaulted the poor boat captain.

After much pain and noise, he agreed to take us out next day, for free. I thought it would be another waste of time, but Gloria wanted her money's worth.

But this day? We had barely left the harbor when miniature dolphins (spinners!) leaped high out of the water, stunning us with their athletic grace.

And then — in the distance, the back of something enormous. It was a gigantic whale — a humpback, the singing whale. It arched in eternal slowness, downward, and it was so close...

"Can I touch it?" I asked the captain, as I pulled on my swim fins and spit into my mask, to keep it from fogging up.

"You can try," he said.

"Be careful, hon!", said Gloria.

I took several deep fast breaths, hyper-ventilating, cramming my system with oxygen, readying myself for one great breath-holding dive.

I slipped off the boat and swam down.

I heard the cry of the whale, was surrounded by that sound, a sonic shiver that embraced and echoed, involving my whole being.

For a moment, I was young again. Every muscle relaxed, I swam down and down, twenty feet, forty feet, even sixty perhaps — toward that beautiful humpback whale. I paused with arms outstretched, completely relaxed, forgetting my need for air.

I saw her entire great length, and the long fins, with their weird little warts.

I saw the tiny eye, and it saw me.

I tried to swim closer, but the whale, with no visible effort, moved back, keeping her distance, in absolute command of her element.

I quit trying to be pushy, just hung there, suspended in beams of underwater sun.

It was the highest moment of my ocean life, I remember thinking; nothing could be better than this —

And then, from underneath this gigantic creature, a baby whale poked out its head... A photographer was there that day, and he sent me this picture.

A quiet moment, involving another Pope, Francis, at the Vatican.

I was not there, at that three-day stem cell research conference. To the best of my knowledge, no one raised the subject of embryonic stem cell research. If it had come up, I am sure the position of the church remained the same. My position also has not changed. I am for it, all the way. But still it was clear that cooperation was sought.

One of the featured speakers was the President of the California Institute for Regenerative Medicine , Dr. Maria Millan.

I hope the leadership of the Catholic Church took notice of her gender.

# 23 The Passing of Gloria

Katherine visits Grandma's grave and uses vinegar to clean the tombstone.

A parade of visitors began. That cheerful bubbling vivacity which drew me to Gloria, also made her friends, and so many! They brought her food and appreciation.

They came to say goodbye.

To each she gave something of herself, a memory, a laugh, a gift, some gentle guidance.

The grandchildren came.

Jason hugged her, the most beautiful warm and tender hug.

Roman Junior came, and made his Grandma chuckle, as only he could.

Katherine, strong enough to help Grandma get up from the couch, sat beside her, listening to her elder's words; she leaned against Gloria's shoulder, and smiled.

Desiree's son Jackson called from Nevada, as did Josh her husband. Gloria asked him to look after me, because he was practical, and I was, well, not.

Roman drove his wheelchair into the living room, and hollered up at Mom on the second floor, when she had moved there. When she could not come downstairs, he sent his phone up to her, a video showing him on TV in his council member job; she took such pride in seeing him in action.

Desiree was here, having flown home to stay with her Mom, taking time off from her job as Athletic Director at the University of Nevada at Las Vegas. She was, I believe, the first Latin-X female Athletic Director in America.

"I love you, Mom," said Desiree.

"I love you," said Gloria, opening her eyes. The words came clear and strong.

I wished she would say "I love you" to me as well, but I feared there would be no more. Of course, we had expressed our feeling thousands of times over the years. But that I might never hear those words again from her?

Desiree helped with Gloria's care.

"No, no, wait, wait!" said Gloria to Desiree one day, to my knowledge the last words she ever spoke. I am not sure what she meant. Perhaps she was objecting to the endless "injections" (rubber tipped soft syringe) of pain-killer and nausea relief, all squirted into her mouth; she hated that.

Or maybe she just loved life so much, and wanted it to continue.

I was in the chair beside the hospital bed all day now, holding her hand, talking to her, saying anything I could think of, but mainly what a great wife and mom and friend she had been through fifty years of marriage — she was my heart.

There was nothing left unsaid.

The hospice ladies were wonderful. They would give Gloria bed baths — I no longer had to hug and lift to get her with me into the shower, always fearful we might fall.

Even so, once we almost dropped her. She wanted to use the restroom, but could no longer support any of her weight. So I took her upper body, thinking the hospice person could carry her legs. But that was a mistake. As if in slow motion, Gloria began to fall.

I shifted her weight, like cradling a baby, and just for second I was carrying her full weight. I was not about to let her fall.

I took her where she wanted to go, and everything was all right.

Our favorite hospice lady, Flora, would always say to Gloria, "Hello, beautiful, we're going to get you all fresh and nice and lovely!"

And then one day she told me what it meant when Gloria started throwing up brown stuff, though she had not eaten anything for days.

Flora cleaned her up. I saw Gloria visibly relax, as if she truly wanted to get all fresh and clean and ready. Ready…

"She is actively dying now," said the hospice nurse. I had been holding Gloria's hand. I squeezed it tighter, told her I loved her, that I would see her in Heaven.

I heard rushing footsteps.

It was Desiree. She had been out running, when she felt the urge to come back home — fast. She had run hard, now she bounded up the stairs, grabbing her Mother's forearm, so we both were touching her.

"I love you, mom", said Desiree.

Gloria breathed a few more times and then her chest stopped rising. It was April 7th, 2020.

With Desiree's help, I negotiated the placement of the grave, near Gloria's Mother Soledad, and other family members, leaving room for me to lie beside my wife one day, when that time comes.

As the great actor Basil Rathbone said, when making similar arrangements for himself and his wife: "Let us be close enough that we can reach out and hold hands, and get up for a walk, if the mood should strike."

Because the funeral would take place during the COVID crisis, the mortuary folk told us, it would have to be a closed coffin ceremony.

I explained to them why that was not going to happen. They changed their minds, and we got to see Gloria one more time.

The funeral went well, if such an event can be so described. Everyone spoke from the heart and gave warmth and honor to her memory.

But instead of trying to remember what everybody said, I want to contribute a poem I wrote for her, and read to her, not long before she passed.

A FINAL VALENTINE?
Not to be sentimental, nor to kick up any fuss,
But there will never be a final Valentine for us;
So long as love exists, and hands are held by girl and boy,
That magic we will always share, our personal moments of joy,
The happiness you gave me, the mischief in your eye,
Adventures we shared, delights for you and I;
So long as story lasts, and woman shall love man,
That long are you with me, always hand in hand.
The times we spent together, relive them in mind's eye,
On our backs on Pennsylvania hillside, smiling at the sky.
Our children's accomplishments will prevail,
Desiree and Roman, a strength that does not fail;
Our grandchildren too, Roman, Jackson, Jason, Katherine the Great,
With laughter and satisfaction, they overfill our plate.
So long as good exists, so long as sky is bright,
That long will you and I go on, overcoming darkness with the light.
With love, your husband, Don C. Reed
And finally, here is what I put in the newspaper.
ALWAYS IN OUR HEARTS

Gloria Reed (6/11/44 — 4/7/2020) passed away April 7th after a struggle with cancer.

Gloria is survived by her husband, local author Don C. Reed, who wrote about her often in his books. The couple had two children, Desiree Reed-Francois, Athletic Director at the University of Nevada at Las Vegas, and Roman Jason Patrick, a stem cell research advocate, as well as four school-age grandchildren, Roman Jr., Jason, Jackson and Katherine.

She contributed to the community for decades as administrative assistant for Char Hawkins in the Special Education department, Fremont Unified School Services.

Gloria loved to dance, and was very graceful on the floor. She always had a smile to greet you with, and was a fantastic cook. Even in her later years, she was a strong athlete and at the age of 70, astonished her grandkids "by sinking fifty-two baskets in a row".

She had goals for her family, and drove husband Don to graduate from Cal State Hayward and become a teacher, her daughter Desiree to graduate from UCLA and earn her Juris Doctorate from the University of Arizona, her son Roman to receive his bachelors' degree from UC Berkeley — and Roman Jr. is close to graduating from Cal Berkeley as well.

An effervescent personality, Gloria enjoyed socializing with her friends in the Red Hat society, as well as her sorority.

A devout Catholic, she worshipped at St. Joseph's in Fremont for many years.

She will be missed by all who had the privilege of knowing her.

I did what was expected of me. But inside, I felt my life was broken, beyond repair. I tried to buck up, reminding myself of what I regarded as her last request: that stem cell research must go on.

Everyone has tears, but cures are what we need.

# 24 The Half-Empty House

I wake to an almost empty house; Gloria is gone. Sometimes I forget, and reach for her in the bed beside me, that spot in which she always slept. But the mattress is cold. Gloria is no longer there, my wife of fifty years, nor will she come again.

Even so: as I go to the bathroom and begin the day's routine, I have a sense of purpose, though my heart still aches with loss.

We in California have something no other generation has ever had: not in the history of the world. We live in a state which can fight back, not just against cancer, but all forms of chronic disease — using stem cells and gene therapy.

California has the opportunity to bring great change: to put an initiative on the ballot to fund stem cell research. In 2004, we did.

And as the old song says: "We did it before — and we can do it again!"

Bob Klein, of course, is the author of Proposition 71, the citizens' initiative which led to the California Institute for Regenerative Medicine (CIRM). He was not spared the pain of loss: his beloved son Jordan died too young, from complications of Type 1 diabetes. Bob's mother also passed away from chronic illness: Alzheimer's disease.

But he brought us closer to the prophesy of Christopher Reeve, who said: "One day, Roman and I will stand up from our wheelchairs, and walk away from them forever." Cure did not come in time for the paralyzed Superman, but the flame of his faith still lights our way; we will "go forward", as he said, and we will prevail.

We cannot know when cure will come; but of this we can be sure: if we do nothing, that is what we should expect — unless, of course, things get worse.

Shakespeare said it best, asking: "...whether 'tis nobler in the mind to endure the slings and arrows of outrageous fortune, or to take arms against a sea of troubles, and, by opposing, end them."

That is the choice, isn't it? Do we fight back against what's wrong, or just give up? If we choose to, we can fight. We can rouse our neighbors, excite the state, renew funding for the California Institute for Regenerative Medicine (CIRM) .

Yes, my wife is gone. I miss her every day, and will for the rest of my life. Does one become accustomed to the absence of a limb?

But Roman is still paralyzed: millions suffer paralysis like him, or cancer, like that which took his mother. Their families also adore them, and want them saved.

And a new horror, COVID-19, is stalking the world.

If you have muddled through this series of 4 books, you have seen the advantages I have had — the lucky breaks, the knowledgeable friends, the accidental gift of a knack with words — well, it is time to put everything to use.

In L. Frank Baum's THE MARVELOUS LAND OF OZ, there is a wonderful scene where everything depends on the outcome of a race between a griffin, inhabited by the spirit of the evil witch Mombi, and a wooden horse, ridden by Glinda, the good witch of the North. And Glinda says:

"Now you shall prove you have a right to be alive! Run — run — run!"[1]

Before us now is the greatest stem cell battle of all time. Every patient advocate must play a part. This is our fight, to win or lose, with everything on the line.

If we win — we gain $5.5 billion in the fight for cures.

If we lose, the program dies.

Proposition 14: the California Stem Cells for Research, Treatments and Cures Act of 2020. It may save millions of lives and billions of dollars.

My house is not completely empty; there is still room for hope.

---

[1] http://classics-illustrated.com"The Marvelous Land of Oz"

# 25 A Science Adventure

N. C. Wyeth's painting of Tom Piccot and the giant squid, whose real-life battle inspired my young adult book, *The Kraken*. (Pinterest)

As an eighth-grade English teacher, I ended every school year by doing a written survey, asking my students what was their favorite — and least favorite — subject.

Their favorite was usually English, which I discounted as politeness on their part.

But their least favorite? That was frightening. Almost invariably, they put science dead last on the list, saying they *hated* it, calling science "useless", "boring", and "having no relation to their lives".

So, every Monday I would do one thing to excite my students about science. Monday was vocabulary day, and I always lectured on the definition of words — but I never told a word without a story, and I always had at least one science story.

Like how a legendary monster was proven to be real — by a fight between a boy and a giant squid, October 17, 1853.

It happened in Portugal Cove, in the province of Newfoundland, off the coast of Canada.

Tom Piccot, his father Theophilous, and their neighbor Daniel Squires, were out fishing on the waters of Portugal Cove, when they saw something floating. It was gigantic, so big they thought at first it might be a dead whale.

Theophilous sank a boathook into the back of the sleeping giant squid — which woke. A forest of suckered arms grasped the small boat and began to pull it under.

The adults, realizing their time had come, shook hands and blessed each other.

But the boy Tom Piccot was not ready to die. He snatched up the bait axe, and attacked the giant squid, chopping off two of its arms. The beast let go, and shot away, to be seen no more.

One of the arms was cut up for dog food. But the other (19 feet, six inches long) was sold to a scientist and preserved; *the first widely accepted proof of the existence of the giant squid.* Anyone who doubted the monster had only to visit the Museum of England.

I read a short account of the attack in Frank Lane's classic book "KINGDOM OF THE OCTOPUS", and was fascinated.

Could it be true?

On our last trip before Roman's accident, Gloria and I flew up to Newfoundland (pronounced Newf-in-land) after sending a letter to the museum, which politely invited us to visit.

"Don Reed's here!" I heard someone shout, as I stood at the entryway to the Newfoundland Museum. I could not believe it — I was a stranger, and they were making a fuss?

John Maunder, curator of the Newfoundland Museum, *unpacked a giant squid* for me. It was thirty-two feet long, five times as long as a man. He showed me how its black beak could shoot out, like a parrot's, but big enough to break a man's arm.

I wanted to meet the descendants of Tom Piccot, so I leafed through the phone book, calling up all up all the Piccos and Piccots (they were casual about spelling in the nineteenth century), until a voice answered: "Oh, aye, that'll be me great-great-grand-da', Tom!"

I met Mark, Lisa and Michael, descendants of Tom Piccot, the boy who fought the giant squid, and saw the ruins of the old stone house in which he had lived.

All around me was science of the most practical variety; nature cannot be ignored in Newfoundland.

For instance, the airlines told us we had only a six-week window in which they could guarantee a snow-free landing. The province was just six hundred miles south of the North Pole. The weather had to be accurately predicted, unless you planned to remain for the season.

Newfoundland is one great rock, with a thin layer of soil over granite. Some of that little bit of dirt was brought in as ballast in the bottoms of boats.

If you fall, it will hurt.

I wondered, what would happen if I injured myself on the visit?

"Oh, we'd take care of you," said our landlady, "Medicine's free here, you know."

Free medical care? Was this possible?

It seemed so.

We stayed at Paulette Kinnaird's Red Lion Lodge, formerly Tom's Piccot's combination church and school; at night we slept under a hand-made quilt, lulled to slumber by the whistling winds of Portugal Cove.

Gloria went fishing with her bare hands ("Don't take any more than you can eat," she was told), gathering up capelin where the small fish schooled.

I knocked on a neighbor's door, asking questions for my book, and Bill Day called up his son, Alex, and asked him to drive me around for the afternoon, so I could "get it right about the lay of the land."

Newfoundland fisherman Alvah Lee showed me how to "face" a cod, removing the facial cartilage for frying; and ninety-two-year-old George Churchill took me for "a dodge along the landwash", to get the details of the cove right.

Science? He told me the fisherman's ultimate horror: the giant factory ships and their nets which scraped along the bottom, taking everything, so the small fish had no chance to grow and become large. And if the nets broke off in a storm, they become "ghost nets", and went on killing, with the fish all rotting in the nets.

And one more thing about Newfoundland, which for me sums up this magic place.

I was in St. John's (locals pronounce it Singe-in) and needed directions for something. I saw a policeman, walking by. I asked my question, and he gave me directions, carefully, even had me repeat them back, to be sure I understood.

But there was something missing... I finally realized what it was, and asked him:

"Excuse me, I can't help noticing — you are a policeman. But you are not wearing a gun — why don't you have a gun?"

He put his head on one side, looked at me oddly, and said:

"Well, you wouldn't want me to be hurting anybody, would you?"

That is what science is to me: visiting, adventures, unexpected questions which lead down strange new pathways;

And that is how I taught it in school: always with people, always in a story: seeking science's connection with our lives.

# 26 Down and Out?

Our beautiful California stem cell program was dying. After sixteen years of carefully managing the research grants, there was barely enough money to pay for its closure. No new proposals were being considered; the California Institute for Regenerative Medicine (CIRM) was essentially done.

A world without CIRM? For we who loved the agency, this was a punch in the gut.

We had come so far! When CIRM began in 2004, the field of regenerative medicine was largely unexplored: stem cell and gene therapy were "terra incognita": the land unknown.

Proposition 71 (with its $3 billion in funding) had forever changed that. From across the country the scientists had come, top-notch veterans and enthusiastic newbies alike, relocating to the Golden State. They applied for grants, as did our "home-grown" folks. To everyone's delight, they did not need to squabble; there was enough to go around, at least for 16 years.

Not only did the scientists work and learn and advance their own knowledge, but *they shared their findings*: more than 3,000 published medical discoveries — a library of advanced information.

This was not just empty talk, theoretical babblings; this was right-now reality: offering the possibility of change, and for the better.

Infant children, fifty of them, who had been diagnosed with the dreaded Bubble Baby disease, (Severe Combined Immunodeficiency, or SCID); normally they would have had little chance; but now thanks to CIRM-funded research, they lived.

Those same techniques, developed by Dr. Donald Kohn of UCLA, worked against a second condition, granulomatous disease, healing a young man named Brenden Whittaker. It might also alleviate (or end!) the agony of sickle cell disease.

And there was more.

Young men and women, paralyzed from the neck down, recovered varying degrees of arm, hand and shoulder motion. This allowed them more control over their lives, instead of requiring 24/7 attendant care. As you know, my son Roman has been paralyzed since 1994. This particular therapy was only for new injuries, so it would not help Roman. But what a step forward!

Blind people had measurable return of vision. 15 patients, all legally blind, (20/200) received a CIRM-funded treatment. Remember, these were people who were almost certainly going to get worse. Instead, one quarter (27%) improved, and 33% stayed the same. Compare that to the untreated people: 80% got worse, and nobody improved. The results were measured on a standard eye chart, with improvement shown by patients seeing more letters on the sight charts — but one woman had a more personal result. She saw the faces of her children, *for the first time*.

With CIRM and stem cell research, change seemed possible. The grim conditions — arthritis, paralysis, heart disease, cancer, Alzheimer's, Huntington's, many more — these were just problems to be solved: difficult but doable.

But if the money was gone? Our hopes for treatments and cures would be put off for years, decades, lifetimes.

On the federal level, the climate for research funding was unreliable, to say the least. The National Institutes of Health is the jewel of America's research system, but it depends on federally approved funding, and that resource has been flat for years, not even matching the increased costs of inflation. The Trump Administration recommended a 20% cut — though this was later rescinded.

A quote from me was used as the tagline for the CIRM Report: "Something better than hope".

"Today, thanks to 7.2 million voters who authorized the California Institute for Regenerative Medicine, we have something better than hope; we have results, accomplishments, people made well — and a systematic way to fight chronic disease." — Don Reed, Vice President, Public Policy, Americans for Cures."

But now the money was gone.

Unless, perhaps, Bob Klein could do it again?

I told Bob once that he was actually a very bad businessman.

He said, "Wha-at?!" Because he is of course highly successful, or he would not have money to donate to stem cell campaigns. But I was just making a point.

"All this work on stem cells, and it never made you a nickel!"

Which was, of course, perfectly true.

As the patient advocate who designed and championed Proposition 71, Bob also had been its largest contributor, donating $3.4 million dollars. Nominated by Governor Arnold Schwarzenegger, and unanimously elected by the board of directors, he served for six years (declining payment) as Chairman of the Board.

Would he take on the challenge another time? He would! And suddenly we were caught up in the sweep of history again.

**Question for Bob Klein:** "When did you decide to renew funding for the California stem cell program? And why so much money? $5.5 billion? Would we not have a better chance of success, if we asked for less?"

Bob's response: "In 2011, I stepped down from the chairmanship of the program. It was clear that the Federal government was not going to allow change in the national policy. The conservative ideologues in Congress could block the needed change. There would be no funding of the most advanced stem cell research, and none for the urgently required clinical trials.

"It was vital that California continue its leadership, so the great discoveries of our scientists would reach the hands of doctors and patients.

"As for the $5.5 billion, it was not an extravagant goal; it was the necessary amount." — Bob Klein, personal communication

I have known Bob since 2003. When he commits to something, it is with everything he has. Whatever comes, he will deal with it.

But there were problems, major ones. The first time around, there had been serious amounts of donor money available, funds to run a campaign. Now, we were headed into a global pandemic, as well as a Presidential election, which tied up many donors' funding. It would be much more difficult, this time around.

Well, nothing for it, but to do it.

# 27 Running in the Dark

An advocate family: Seth and Lauren Rogen, and her brother Danny, united in the fight against Alzheimer's disease.

At 5:30 in the morning, I went for a run, carrying my big Tai Chi stick, in case I was aggressed by dogs. I had learned two patterned sets with the staff, and it actually worked in a fight. I had once flicked a snarling dog, and remembered yet how the energy vibrated up the staff and rapped the animal on its slavering jaw; it did no damage to him, but utterly changed the situation. I hoped there would be no savage turkeys or threatening skunks, both of which I had encountered.

To the left of the trail was an almost dry channel: a trickle of water from the Fremont hills; on my right the backsides of darkened houses.

The red cloth of my mask flapped in and out against nose and lips. I was not likely to meet any COVID folks at this hour, but things were a little strange. The other night a helicopter had swooped down near a gaggle of folk, politely asking us to disperse; it sounded like the voice of God.

A sliver of moon illumined my path. I heard the shuffle of my own footsteps, as I turned between the trees, toward my street.

As always, I thought about Gloria, rehearsing how I would tell her, make a story out of this little moment. Then, of course, I remembered.

Some days were hard; yesterday I kept finding myself wandering around the house, like that dog at a funeral in the movie SHANE. It kept pawing at the coffin, not understanding why the master was inside.

I tried to focus on what I could control, what tiny chore might help pass the $5.5 billion dollar renewal of CIRM. I wrote and posted another essay, about how it was not too late to get signatures. We did not want to miss qualifying for the ballot!

People helped. My next door neighbor gathered 5 signatures, Mimi Gardner from the office got two more, my sister Barbara did another 5, my brother and sister-in-law contributed two.

Fremont Assemblyman John Dutra's son Dominic had kindly brought over a dinner for me, and I asked if he would sign a petition? He would and he did.

Two visitors stopped by and I volunteered them as well. I dashed upstairs, printed out the required 16 pages, got a 9x12 envelope and two forever stamps — ran back down, helped them fill out the form, stamped the envelope for them, made sure they knew where the mailbox on the next block was — "Wait for the clunk," I told them, "Be sure it falls in," and they promised.

How does great change happen? For me, it was minutiae (great word, huh?) tiny steps, like oceanic plankton, almost invisibly small, but which feeds whales.

Every morning I scribble, in the afternoon I study, getting ready for tomorrow.

For the group, success depended on endless communication. Twice a week, Monday and Thursday, Bob would call a ZOOM meeting, to discuss and encourage the campaign. No matter the subject, he would always have a way to improve it.

The people I worked with most? More on them later, but briefly:

Melissa King was our endlessly energetic leader of field operations. She had not only been deeply involved in the first stem cell campaign, Proposition 71, but had held an important position at CIRM.

Bob Klein's son Robert could be found wherever the work was thickest. Like his father, he lived in two worlds: stem cell research funding, and the real estate efforts that made the campaigns possible.

An unsung heroine was Elizabeth Tafeen, Bob's right-hand person. She has an amazing memory (eidetic?) appearing to remember literally everything.

Jacqueline Hantgan, returnee from the first campaign, was our outreach specialist, contacting groups like Hadassah and NOW, explaining to them why endorsing the stem cell initiative, Prop 14, was absolutely in the best interest of their membership. We ended up with more than 80 such endorsements, which can be seen at our website, www.caforcures.com

Movie star Seth Rogen and his wife Lauren were advocates against Alzheimer's disease. Lauren's Mom died of the condition. Lauren is dedicated to raising funds for a group, HILARITY FOR CHARITY; she is a recent member of CIRM's board of directors. She and I wrote an op-ed together. And Seth? He was the voice of "Stemmy the Stem Cell:", our cartoon spokes-character.

Anna Maybach had worked with terminally ill children at Vanderbilt Children's hospital, where she wrote a paper on the importance of involving the children in their own decisions; too often, she said, they were routinely left out of the discussion. She worked on content creation for the campaign — and had been scuba diving in the ocean since age 12!

UCLA professor Mitra Hooshmand, Science Director for our campaign and advisory board, was not only brilliant in science, but had magnificent eyebrows (sorry, Mitra, but it's true!) an infectious laugh, and a reputation for never turning down a chore.

Senator Art Torres (Ret.,) had rare personal charm. He was like the legendary frontiersman Davy Crockett, whose smile could reportedly grin a squirrel off a tree. I don't know about the squirrel, but the Senator did bring an incredible number of friends to the table. If somebody said, "what about Senator X?", Art would invariably say, "Let me give him a call," and good things would follow.

And of course, there was Bob, like Mount Everest in the back yard.

Our campaign was united by love of CIRM. Whatever the disease we personally fought — paralysis, cancer, diabetes — CIRM was fighting it too.

Practicalities: to get on the ballot, we needed 623, 212 valid signatures. Bob set a goal of a million, to be sure we had enough, in case some were disqualified. Disqualification of even one signature could be more dangerous than at first appeared. If you made a mistake, — there might be penalties as well. San Diego, for instance, would "fine" you *32 times the number of duplicates*. Three invalid signatures meant subtracting 96 from your count!

Before COVID struck, we planned to do what worked before; go where the crowds were, set up a table, rush out and tackle folks, get those signatures. We could work in front of a library, or (our favorite) at Fisherman's Wharf, in San Francisco.

But with the pandemic? Crowds were illegal. Cities were like ghost towns.

So, we gathered our signatures one by one, with the novel approach of the "Sign at Home" Program, where people could print a petition emailed to them by our coalition members or campaign team, sign it, and send it in.

With our COVID masks on, we walked up to neighborhood houses, rang their doorbells, offered our clipboards.

Everything came down to the signatures, do-or-die for the initiative. With them, we had a chance to put our case before the voters. Without them, it was over.

It was painfully slow, and we were under a severe time crunch. If not enough signatures were gathered quickly, a full count (instead of the random sample) would be triggered, and the counties might not have time to finish sampling by the deadline. If it sounds complicated, that is only because it was!

Those who liked what we were doing? They wanted to talk — we didn't have time, and had to be politely short, almost rude, leaving quickly, to try for more.

One lady asked me in, then raised a dozen objections, plainly having studied the initiative. It was like talking to a polite lawyer. Finally, I said perhaps we should agree to disagree, and got up to go.

"But don't I get to sign the petition?", she said.

One person at a time, one signature sheet at a time, one emailed group at a time.

If we could not do them fast, we must do them slow — but keep on going.

Familiar faces and new worked with us on this campaign: veterans from the first campaign, like James Harrison, who, once again, played a central role in developing and finalizing the language of the initiative as well as providing essential counsel throughout the campaign.

Steve Merksamer and the team at Nielsen Merksamer once again, as they had in 2004, managed the operations of the C4 organization and campaign committee, and managed the bank account and finances for the campaign — always an important job.

Paul Mandabach, a veteran of 2004's Prop 71 as well, brought his team, described by Brett Noble:

"Winner & Mandabach Campaigns served as the lead campaign consultant for the YES on 14 campaign, as they had done previously for the Prop 71 campaign in 2004. The firm's President & Managing Partner, Paul Mandabach, oversaw strategy and implementation. Vice President Brett Noble served as the day-to-day account manager, and partner Adam Stoll contributed to strategy and paid media. Winnner & Mandabach veteran Bob Deis also provided valuable insights on measure design, advertising creative development, polling, and overall strategy.

"BASK Digital Media served as the digital marketing arm for the YES on 14 campaign, handling website development, digital advertising, social media efforts, and other digital activities. BASK was led by Amanda Malo and Andrea Grohovsky, who spearheaded the team's digital strategies and day-to-day digital efforts. Alexi Melssen and Paige Severson managed digital advertising planning and placement. An all-hands-on-deck team also included Rachael Kelch, Anna Blaszkiw, Madelaine Baird, and Andrew Bartz.

Fairbanks, Maslin, Maullin, Metz & Associates (FM3) served as the lead pollster for the YES on 14 campaign, as they did previously for YES on 71, overseeing opinion research activities related to design of the measure, message refinement, ballot argument drafting, tracking surveys, and focus groups. FM3 partner Dave Metz let the efforts for FM3, along with Vice President Lucia Del Puppo.

2004 veteran Fiona Hutton returned in 2020, this time with her own namesake PR company. Hutton's team was led by Kendall Klinger, who served as the campaign's Communications Director. Dagny Ellenberg worked with the campaign team on the final months of coalition building and mobilization, and Sarah Melbostad managed media relations. Their team was supported by a talented bench of storywriters, earned media experts and outreach specialists, including: John Koriel, Amanda Bobbitt, Sierra Layton, Evan Swerdfeger and Melanie Tuberman."

The firm Amplified Strategies participated in the YES on 14 campaign as part of the research and data analysis team, providing detailed microtargeting for digital advertising placement. Amplified's effort was led by Andrew Meyers with contributions from Mike Kamman and Mike Meyers."

The troops were gathered.

# 28 R U Latin-X?

And this is how I remember Gloria — in a beam of sunlight, in the doorway of a church — and I almost hear the rustle of her lace wedding dress.

For half a century, it was my privilege to be married to Gloria, a Latin-X woman. This not only gave me joy, but also some awareness of her ethnic group's medical needs.

If you, like Gloria, are of Latin-X decent, you face six disproportionately deadly threats: diabetes, heart disease, obesity, kidney disease, liver disease and cancer.*

---

* https://www.nm.org/healthbeat/healthy-tips/common-hispanic-health-issues

Gloria shared my stem cell life, as here with CIRM scientist Pat Olson, on the day my first stem cell book was released. (blog.cirm.ca.gov)

What do those six conditions have in common, besides their huge impact on the Latin-X community?

All of them are being fought by the California Institute for Regenerative Medicine (CIRM), our stem cell program.

To check with the experts, just go to www.cirm.ca.gov and look up each disease. Or, visit some of the URLs below.

Let's look, very briefly, at one possible CIRM-sponsored therapy for each.

1.  Diabetes: a pouch full of specially prepared stem cells will be put under the skin; there it will take in nutrients and give back insulin — while being protected from rejection by the body;[1]
2.  Heart Disease: stem cells will be mounted on a biodegradable scaffold, a sort of living tire-patch, to cover heart attack scars and replace them with healthy tissue;[2]

---

[1] https://care.diabetesjournals.org/content/early/2018/03/20/dci18-0007
[2] https://www.cirm.ca.gov/our-progress/disease-information/heart-disease-fact-sheet

3. Obesity: the color of your fat impacts your health; brown fat burns off rapidly, white fat clings as blubber. There might be a genetic fix to change fat from white to brown, from harmful to helpful...[3]
4. Kidney disease: for those who must endure the blood-washing of dialysis, a tube has been invented that can stay partially inside the body, joining with it on a cell by cell basis, reducing the chances of infection;[4]
5. Liver disease: when the liver fails, death is near; a transplant will help, but there are very few extra livers available for donation. Instead, it may be possible to grow and transplant new and healthy liver *cells*, instead;[5]
6. Cancer: There is a cancer stem cell (like an evil twin) which the body cannot detect; it has a protein that makes it invisible to the body's defense mechanisms (macrophages). But if we can strip away that protein protection, the macrophages will hopefully detect and eat *all* the cancer.[6]

Many of the above therapies (and more!) have shown early but promising results — and possibilities.

Do you see why I love CIRM — and why it must be continued?

---

[3] https://tinyurl.com/y64edfcg
[4] https://www.cirm.ca.gov/our-progress/disease-information/kidney-disease-fact-
[5] https://www.cirm.ca.gov/public-web-disease-focus/liver-disease
[6] https://www.cirm.ca.gov/our-progress/disease-information/solid-tumor-fact-sheet

# 29 The Multiplicity Disease

Our daughter Desiree Reed is Athletic Director for Missouri University — yay, Mizzou! (www.showmemisssouri.edu)

Imagine a disease so deadly that it would attack almost every system in your body. Further, this vile condition would attack the two people you love most in the world: your baby, and his/her mother.

Many years ago, when our first child was about to be born, I planned to be in the delivery room.

I had read books about the Lamaze method of natural childbirth,[1] which said the husband should be part of the birthing process. That made sense to us.

We practiced breathing techniques every night to make birthing easier. She was the athlete; I was the coach.

---

[1] Painless Childbirth: the Lamaze Method, by Fernand Lamaze

Mana Parast is fighting to cure preeclampsia, which threatens both mother and child. (providers.ucsd.edu)

I was not entirely sure what to expect. We were shown a medical movie; it turned my forehead cold and my stomach did strange things. I had planned on comforting Gloria, but she said, "Oh, this is interesting!" I found myself sinking down into my chair, while she patted my hand and told me everything would be all right.

And one of my friends warned me: whatever you do, don't look at the placenta, it will gross you out.

In the delivery room, Gloria was a star. She strained her muscles and shouted like an athlete. I stood at her shoulders, whispering encouragement, mainly trying to stay out of the way. Back then, a husband's presence in the delivery room was not usual, and the doctor had made it clear I would be ejected if I caused any problems, i.e., fainting.

But Gloria was busy and the staff soon ignored me, yelling code words to each other, like "crowning!" which apparently meant "I can see the top of her head!"

Suddenly, the doctor put a baby in my arms, Desiree Don Reed, middle named in pride after me. Some say newborns are ugly — maybe theirs are — but mine was perfect in every way. She did not cry. In the most dignified manner, she surveyed her surroundings. She looked at me, and smiled. She looked at her fingers, closing them into a fist, like hmm, what is this?

The nurse took her from me, and I felt like growling. But then an intern handed me some lime gelatin, saying Gloria could eat something now. I tried to feed her, but she was too exhausted to chew. The little green chunks fell out of her mouth.

I asked the doctor, where was the placenta? He gestured with his head. And there it was, on a corner of the bed: just a limp torn plastic-looking white bag: a few streaks of blood, but nothing gross or shocking. What I felt for it was...gratitude. In my mind I actually said "thank you" to the placenta, for sheltering my daughter inside the womb, those nine long months.

It never occurred to me both Gloria and Desiree might be at risk. We did not know about any disease we had dodged. We were just two twentysomethings, at the hospital to get our baby.

But what if there had been something wrong with the placenta, that protective connection between mother and unborn child?

Like pre-eclampsia...

"Pre-eclampsia is a pregnancy complication...(which threatens) 5–8% of all pregnancies. It has major effects on blood pressure and kidney function of the mother (and) is responsible for a significant proportion of maternal deaths and growth restricted babies ..."[2]

A living nightmare: "These babies are at increased risk for long-term disabilities, including cerebral palsy, gastrointestinal problems, vision and hearing loss...(and) increased cardiovascular disease and diabetes in adulthood."

Desiree could have been born deaf and blind, prematurely...

"On an average day in California, 149 babies are born prematurely. Many of these babies will require weeks of care in an...intensive care unit... at an average cost of $125,000...compared to $1,500 (for) a baby born at (normal) term..."

Dr. Mana Parast, a CIRM-funded scientist of the University of California at San Diego, is trying to reduce that suffering, and expense.

A difficult challenge. For one thing, the usual test subjects, rats and mice, do not get preeclampsia the same way people do.

The main problem in preeclampsia is a shortage of *"trophoblasts"* a vital stem cell. There are two kinds: syncytiotrophoblasts, (pronounced like sin-city-o-tro-pho-blasts) which pass nourishment from mother to unborn baby; and invasive trophoblasts, which help (give) maternal blood to the placenta..." — Dr. Mana Parast, personal communication.

---

[2] https://www.cirm.ca.gov/ourprogress/awards/humanpluripotentstemcell-basedtherapeuticspreeclampsia

Dr, Parast has made one stem cell model of diseased trophoblasts , and another for healthy ones; she may be able to use the healthy stem cells to fight the sick ones — or test new medications on the stem cells, outside the body.

"(Her) work focuses on the stem cells that give rise to the placenta... Her lab's stem cell model — a first — (may) be used to study stages of placental development.

The CIRM grant "will fund continuation of this work...to identify potential stem cell-based therapies for treating preeclampsia..."[3]

A valuable project!

But... Dr. Parast is young. what if she had not been able to get a grant?

"According to NIH data, the average age at which PhD scientists receive their first R01 grant has increased from 35.7 in 1980 to 43 in 2016. And now, for MD investigators, the **average age is 45 years old.**"[4]

When Dr. Parast tried for her first grant, she was 35. If the California Institute for Regenerative Medicine (CIRM) had not been there, she might not have found funding for her stem cell research. She might have had to leave the field.

Fortunately, that need was foreseen by Bob Klein.

I asked him, what was the most important grant? He answered immediately.

"The grants to help young scientists to survive financially", he said, "while they are getting started in the stem cell field."

Across the country, research funds are increasingly difficult to find. The primary source of medical research funding, the National Institutes of Health (NIH) has been flat for a long time — not even adjusting for inflation — and the Trump Administration nearly cut its budget by 20% — twenty per cent! Although the crippling subtractions were later reversed, still there needed to be more.[5]

---

[3] https://health.ucsd.edu/news/releases/Pages/20121212parastawardedCIRM-grant. Aspx

[4] — https://www.aamc.org/news-insights/nih-research-community-target-next-generation-scientists

[5] https://www.theatlantic.com/politics/archive/2017/03/trumps-budget-cuts-nih-funding-by-20-percent/519771/

CIRM established 4 kinds of grants aimed at just such early-career researchers:

1. New Faculty grants I (which 22 investigators received);
2. New Faculty II (23 recipients);
3. The Physician Scientist Translational Research Awards (15)
4. — and the Medical College Repayment Award (5).

Source: Dr. Patricia Olson, chief scientist, CIRM, personal communication.

Without CIRM's help, those 65 scientists might have been lost to the field. Instead, they are working today to fight chronic disease: to ease suffering and save lives.

Dr. Arlene Chiu, CIRM's former Chief Scientific Officer, remembers: "Those grants were designed to encourage (new stem cell) investigators… providing salary and research funding for up to five years, ensuring stable, secure financial support as they begin their … scientific careers."[6]

Providing financial stability for new scientists is one of many reasons the California stem cell program was made.

And why it must not die.

---

[6] https://www.cirm.ca.gov/aboutcirm/newsroom/pressreleases/06282007/stem-cellinstitutesolicitsnewfacultyawardproposal

# 30 Fighting Forward

Bob Klein, CIRM's Chair Emeritus, meaning forever. (blog.cirm.ca.gov)

After Gloria was diagnosed with cancer, I had dreams of fighting. Something clutched my legs and I tried to kick free — but my blankets clung tight — and the eagles were coming! I had a garbage can lid and was fighting to protect my family, but the eagles coalesced into one evil monster bird. I grabbed the creature and flipped it over my shoulder, so hard I also <u>threw myself off the bed</u>.

There was an orange plastic garbage can beside the bed, and my crashing weight exploded it. Shards of plastic stabbed me: face and forehead, neck and chest, dark liquid seeping.

I said "Hon, I am wounded! I'm bleeding!" and Gloria said, "Get a band-aid."

"No, no! I was fighting in my dreams, see, and — ".

Brigitte Gomperts' expertise on lung reconstruction may be of inestimable value in repairing the physical damage of COVID-19. (Stemcell.ucla.edu)

"That's nice, dear, I hope you won," said Gloria, and went back to sleep. I went to the bathroom and plucked out the plastic pieces — band-aids all over the place!

My grand-daughter Katherine came over the next day, and I told her the story, whimpering bravely.

Katherine folded her arms, looked at Gloria, tapped her foot and said, "And where were *you* when all this was happening?"

Yeah, you tell her, Katherine!

Then she got a cloth and hydrogen-peroxide and violently scrubbed my face.

I put a pile of pillows on the floor beside the bed, and they are there to this day.

Meanwhile, behind the scenes at campaign headquarters...

There had been discussions about trying a <u>different date</u> to renew the initiative — maybe the 2022 elections?

But a Gallup poll showed how Democrats and independents turned out in greater numbers (as much as 20% higher turnout) in Presidential elections: they were our strongest allies: whereby Bob chose 2020.

He spoke of the importance of "reporting back to the public," meaning to share the accomplishments of CIRM. And this was vital.

People knew about stem cells in a generally positive way. They gathered it was a good thing. The research had strong support among Democrats, and Independents, even a near majority (42%) of Republicans. GOP leadership opposed us, officially, but the regular folks just wanted cures and health like everybody else.

A major poll (Harris/Interactive, 2010) showed 73% of the public in support.[1]

But hardly anyone remembered Proposition 71, which had, with the research funds it provided, made California the stem cell capital of the world.

I learned to ask people if they knew about the movie, "THE BOY IN THE PLASTIC BUBBLE", with John Travolta. They would almost always say yes. (Apparently a *lot* of people saw that movie!) I would tell them how the California stem cell program had a cure for the real-life "Bubble Baby" disease in that movie, technically called Severe Combined Immune Deficiency (SCID). Fifty children's lives had been saved, and the same procedure might help fight sickle cell disease.

But they shook their heads at the price tag. $5.5 <u>billion</u>?

Way too much, we were told. Ask for less! Wait for better times! Look at the economy, in such bad shape, and the environment: we never had so many forest fires, and the high levels of homelessness, the cost of houses and medical care — we couldn't afford to spend so much money for stem cell research — not right now!

But, we argued, this was actually the best time to renew the funding, so we would not lose momentum in the progress toward cures.

More than ninety clinical trials (now over 100!) were testing new therapies for varieties of cancer, paralysis, arthritis, bone disease, COVID, kidney failure, ALS, leukemia, heart disease, and more.

Right now, we have 71 therapies in trials which CIRM directly helped fund — and an additional 30 human trials that CIRM helped fund in their early tests.

You can check this out in two places.

https://www.cirm.ca.gov/clinical-trials

and —

https://www.cirm.ca.gov/our-progress/clinical-trials-based-cirm-grants.

It would be a huge mistake to walk away from that progress now. Jobless scientists would be scattered across the country; it would be hard to get them together again. The best would be snatched up immediately, accepting other work — the young ones might leave the field altogether, unable to find employment.

---

[1] https://www.cirm.ca.gov/blog/10072010/new-poll-finds-widespread-support-stem-cell-research

And the lab animals — what would happen to them? They might have to be put down. That bothered me. Lab animals lived and died to benefit humanity; they deserved respect, care, and whatever small pleasures they could be provided. I had seen scientists take their lab rats out of their cages and play with them, systematically, to make sure each one had a little bit of fun. But if there was no money?

CIRM fought on with the research, doing what had to be done, step by step — - as the dollars slipped away. We were not completely broke, as I had previously thought, but close to it, and there was no new money coming in.

For us in the trenches, COVID-required Zoom meetings with Bob kept us focused. We shared what worked, and what didn't. We wrote in our priorities ( the top three goals) every week, asked for assistance when we needed it.

I had to work with an editor, something I had never done with stem cell writing before. I had been an editor myself, working for a magazine, and of course all my books were lightly edited by the publisher. I did not see the need to have an editor for op-eds. But Bob was a lawyer, and thought like one. Perhaps he worried that one incautious remark on my part could bring trouble.

So now I had editors, Anna Maybach, and Robert Klein (the younger) Bob's son. Rob was more conceptual, and would discuss changes from a philosophical angle. Anna went through my work with a meat-axe. The fact that she was generally correct did not help.

The big newspapers were approached individually, by Fiona Hutton and Kendall Klinger, Fiona's PR company being hired in early 2020 for that purpose.

I went after the smaller papers, the locals: sending op-eds and letters to the editors, over a hundred, and each one had to be different, an individual submission. A few I got, most not. But each one put our case before the editors. In a small paper, the op-ed editor was often the owner; talking to him/her was a privilege.

And what was CIRM doing to advance their case politically? Nothing. It couldn't. By law, a government agency is not allowed to advertise for itself.

All they could do was keep fighting for cures — and that they never stopped doing.

It was frustrating! When COVID reared its ugly head, CIRM was essentially broke. At a time when a research-funding agency would have been most valuable, ours had almost no money. They had done nothing wrong, been careful with every dollar, but the program had just run dry.

What could be done against COVID? CIRM scraped together emergency funding, $5 million: dividing it among 17 projects, in small amounts. Instead of $3 million or $6 million per grant, now they were lucky to get $200,000: seed money.

Scientists scrambled to see if their projects might have a COVID-19 application.

Some, like Brigitte Gompers of UCLA, had been working on lung disease[3] for years; her expertise fit the need immediately. But the grant she received, for using organoids to test high numbers of drugs very quickly — got only $150,000 ($149,998) though clearly deserving of much more.

And always the aching question, which would not go away: what if there was no CIRM?

---

[2.] https://tinyurl.com/yyphdfub
[3] https://blog.cirm.ca.gov/tag/dr-brigitte-gomperts/

# 31 Against the Plague

Talking to Bob, everything big seemed so natural: like *of course* we should go for a program costing five (and a half!) billion dollars. Even the half was five hundred million bucks!

Bob adjusted his answers to the size of the problem. If a problem was big, so must be its solution.

The California stem cell program would be paid for by the sale of government bonds. There would be no payments at all for the first five years. After that the payments (which would come from the general fund) would be spread out over about thirty years. It would cost every Californian roughly $5 per year: the price, as Bob put it, "of a bottle of aspirin".

There would also be more financial benefits coming to us: the add-on grants (additional funding from other sources) were virtually certain. For the first instalment of the program, Prop 71 add-ons brought $3 billion, as much was spent on the research program. Also, a University of Southern California (USC) economic study revealed an increase in state economic activity of roughly $10.3 billion[2] — overall, there was no question we would get our money's worth, and more.

Besides, consider the alternative: to continue paying endlessly for chronic disease — never stopping, no end in sight?

Ninety per cent of all American medical costs are from chronic conditions, long-lasting (or incurable) disease or disability. That annual healthcare spend in the US is roughly $3 trillion dollars — without curing anybody, just maintaining them in their misery.

---

[1] Capitolweekly.net/stem-cell-initiative-save-lives-and-energize-the-economy/
[2] https://www.cirm.ca.gov/about-cirm/newsroom/press-releases/10092019/
new-report-says-stem-cell-agency-produces-measurable-0

For example: diabetes, every year, costs America about $300 **billion** ($327,000,000,000) more than sixty times the the original California stem cell program provided for research.[3]

The plague we faced was not just COVID-19, horrific as that was. Chronic disease, like a multi-tentacled monster, attacked us through many conditions: heart disease, cancer, paralysis, Lou Gehrig's disease, Alzheimer's, multiple sclerosis, on and on.

More than half of all Americans (51.8%) have one or more chronic diseases or conditions.[4]

California wanted to heal people, to make them well: not better, but well.

First, the program had to be the absolute best it could be. For me, it already was. But others disagreed, including Bob. There were changes to be considered.

Meetings were held, up and down the state, a "listening tour" organized largely by Melissa King, Executive Director of Americans for Cures, the Bob Klein-chaired organization laying the groundwork for the 2020 campaign, to gather suggestions. There were many sensible ideas for improvement. Here are six:

1. Bob wanted the governing board to represent all of our widespread state, not just some areas. Currently underrepresented was the great "Inland Empire"of Riverside, San Bernardino and Ontario, an area big as ten small U.S. states — 27,000 square miles. This could be changed by adding board members from those areas.

2. There should be 2 nurses with real-life experience administering clinical trials — make the board bigger, to have room for Florence Nightingale's kin. This we learned from our friends in the UC Davis stem cell program, who pointed us toward a nurse training program specifically for clinical trial training, developed and run at City of Hope, to learn more about this need.

3. Training grants were needed, so new scientists could learn how to work with stem cells and potential stem cell treatments.

4. Shared labs? Scientists needed the best equipment, but it was so expensive — why not share? A super-microscope, to track the

---

[3] https://www.diabetes.org/resources/statistics/cost-diabetes
[4] https://www.cdc.gov/pcd/issues/2020/20_0130.htm

movements of a single molecule? Good idea, but set up a schedule, let several groups of scientists have a turn.

5. And how about a guaranteed $1.5 billion dedicated to brain and central nervous system conditions, like Alzheimer's, Huntington's, Parkinson's, stroke, *spinal cord injury*...the words made me shiver. My son had been paralyzed 26 years — was there a chance he might be cured?

6. Perhaps most spectacular of all, a 15-member committee would be charged with finding *ways to help clinical trial patients gain access to the new therapies even if they would have normally not been able to afford it...*

I loved that part, because it seemed to me that if costs could be managed to be widely inclusive in the trials, they could be similarly managed in the real world too. It would do no good, by my way of thinking, to have a cure nobody could afford!

But still we had one enormous problem, attacking the nation, and the world.

When COVID shut down the nation's businesses and schools, it also made signature-gathering extremely difficult.

To get on the ballot, we needed 623,212 valid signatures.

We also needed more: a cushion of signatures, in case some were invalidated. Bob wanted at least a million; we must not be stopped a few signatures short.

It was difficult now, gathering signatures. With Prop 71, we worked the crowds.

But now there were no crowds. The streets were empty, the stores were shut, all was hushed and still.

I remembered that line about the plague from Edgar Allan Poe:

"And darkness, decay, and the Red Death held illimitable dominion over all."

---

[1] https://www.census.gov/quickfacts/CA

# 32 Of Tigers, Pain and Sickle Cell

One of the earliest (and best) nature writers of all time, James Corbett; see *Man-eaters of Kumaon*, Amazon.com.

"There is no luxury like the sudden relief from pain." — James T. Corbett

A great hunter, James T. Corbett, was once asked to relieve a village in India of a horrifying scourge: a man-eating tiger. The animal had stepped on a porcupine quill, which broke off inside its foot, rendering it unable to obtain its normal food.

Advocate Adrienne Shapiro and daughter Marissa, battling sickle cell anemia. (blog.cirm. ca.gov)

Dr. Don Kohn is using his remedy for one disease to fight another: challenging sickle cell disease with a gene therapy developed for "bubble baby" syndrome. (Kohn/UCLA's Broad Stem Cell Center, stemcell.ucla.edu)

And so, the tiger went up on a hill and studied the villagers, apparently identifying the weak, which it then began to hunt, systematically, as a food source.

It killed and ate 461 people.

The services of Colonel Corbett were sought, and obtained. He brought his gun and set up, high in the branches of a tree. But he had an abscess of the eardrum, which (being close to the brain) had him in agony until midnight, when the drum burst. Blood ran down the side of his head, and the pain quit instantly.

"There is no greater luxury," he later said, "than the sudden relief from pain."

When the tiger arrived in the moonlight, looking up at the hunter in the tree, Corbett killed it, ending the assaults on the village.

Recommended: "MANEATERS OF KUMAON, AND OTHER STORIES", by James T. Corbett. Corbett is a great writer and environmentalist. You will like him.

African-Americans, Hispanics and Native Americans know about pain: from the sickle cell disease (SCD), which so viciously affects them.

A genetic mutation produces red blood cells that (instead of being round and soft) are C-shaped and hard, plugging up the joints, gouging into the vein walls. An SCD sufferer has constant low-grade pain, and every so often there will be a "crisis" in which the pain becomes excruciating, requiring hospitalization.

Adding insult to injury, when the crisis occurs, the pain becomes so great it may be misinterpreted as drug use; and the patient may be denied proper pain medication.

Imagine going to the hospital in terrible pain, but the nurse or doctor won't help because they think you are a druggie? Sadly, patient advocates inform me, this prejudicial diagnosis is not uncommon.

Children with the condition often have strokes by age ten, damaging the liver, heart, lungs and brain — and the average life expectancy for an SCD sufferer? Just 36 years.[1]

This must end.

One of our greatest science champions is Dr. Donald Kohn of UCLA. He is the person responsible for curing fifty "bubble baby" children:

---

[1] https://www.cirm.ca.gov/clinical-trial/clinical-trial-stem-cell-gene-therapy-sickle-cell-disease

sufferers of a generally fatal disease: with the gigantic name of: adenosine deaminase deficient severe combined immune deficiency — ADA-SCID disease.[2]

When the body has no functioning immune system, a common cold can be fatal.

The first to be cured of "bubble baby" disease was Evie Vaccaro (her amazing story is told in my previous book, "CALIFORNIA CURES") who was diagnosed as an infant.

To save her life, Dr. Kohn removed blood-forming stem cells from her bone marrow, and replaced the trouble-making gene with a newly functioning ADA gene, using an inactivated virus as a vehicle to get it there. Evie is alive and healthy to this day

That technique — which I call the "R-3" technique — Remove, Repair, Replace — may work with other conditions.

That is what Donald Kohn is working on today, trying to adapt the technique to fight Sickle Cell Disease.

I interviewed Dr. Kohn recently. I wanted to know: would the technique to heal ADA-SCID be the same one applied to Sickle Cell patients?

"Pretty much", he said, "There are differences, but the basic principles are the same. We have treated two patients thus far and hope to enroll 4 more in this first Phase I trial.

"We opened up the first version of the clinical trial of the gene therapy for sickle cell disease in 2015... we treated one patient with our first version of the gene transfer approach but couldn't get in enough cells. Unfortunately, the treatment was not successful for her.

"So we spent several more years in the lab to improve the gene transfer approach... we now have FDA approval to make that change in the protocol...."

Might it work with other diseases as well?

"About 40–50 of them," he said, "other immune deficiencies, and blood diseases... especially those characterized by a build-up of toxic materials in the body's cells, as a result of enzyme deficiencies, which may affect... the skeleton, brain, skin, heart, and central nervous system."

How did he get started in this field?

"I was a nerd in High School," he said, "I found science fascinating. By 1985, I was already studying immune deficiency, and stuck with it, ever since."

---

[2] https://rarediseases.org/rare-diseases/lysosomal-storage-disorders/

What did he think about the California Institute for Regenerative Medicine?

"I am a huge fan of CIRM," he said, "CIRM helps in all three phases of research: Basic, where we study and search, Translational, trying to develop a usable therapy, and Clinical, where we test it out."

Were there any special needs he had as a scientist, right now?

" *Lab technicians* do not have financial security," he said, "It would be helpful if there was a grant specifically for them."

Did he have children?

"Yes, two. One is a pediatric immunologist, and the other is a computer scientist — he is probably the really smart one."

Wait a minute. Dr. Kohn is so smart I can barely follow his conversation — one of his children is a pediatric immunologist — and his *other* child is the smart one?

"He's into computer games," he explained.

Dr. Kohn is not working alone. CIRM is in a partnership with the National Heart, Lung and Blood Institute, with the National Institutes of Health.

The partnership's title is simple, and strong; it is the "Cure Sickle Cell" initiative.

CIRM is already fighting hard.

To date, CIRM has given sickle cell disease grants totaling $51,344,261.00. Fifty-one million to support the sickle cell efforts of such outstanding scientists as:

Joseph Rosenthal of the City of Hope;

Mark Walters of UCSF;

Pierre Caudrelier of ExCellThera, Inc.;

David Williams of Boston Children's Hospital;

Matthew Porteus of Stanford University, and others.

Their varying approaches can be read at "Sickle Cell Disease Fact Sheet: California's Stem Cell Agency".[3]

May their funding continue — and may they bring relief to all who suffer!

---

[3] https://www.cirm.ca.gov/our-progress/disease-information/sickle-cell-disease-fact-sheet

# 33 How Many Chronic Diseases Do You Have?

If I may ask a personal question — how many chronic diseases do you have?

I have four myself, but don't worry, they are not contagious — especially not through the printed page! I have:

**Arthritis**: of the hip, knee, shoulder and back. Mostly, it is just the normal wear and tear of age.

But the back injury was brought on stupidity. I carried a 300-pound pump up a flight of stairs — instead of waiting for help — and lowered it down the other side. I felt a soft *pop* as the vertebra collapsed. The pain was sickening. I fell to my side and let the world go on without my help for a while.

"C 5,6 compression fracture," the doctor said, "You will have good days and bad ones." He was not wrong. I seldom bound up the stairs anymore, and even rising from the floor is not altogether graceful. Some days the pain is troublesome, but I work around it.

**Peripheral neuropathy (PN)** in feet, ankles and calves. This pins-and-needles nerve condition wants to spread, but is manageable with medication. It feels like numbingly-tight socks all the time, even when I am barefoot. If I forget the medicine, pain will remind me. It can be like eagle claws, sinking in. I don't forget often.

What happens is my nerves do not properly re-insulate: a fatty acid called myelin is supposed to wrap itself around each nerve. Mine does not replenish itself as it should. (Peripheral means on the edges: neuropathy means nerve damage). It can be caused by injury, diabetes, old age, paralysis (Roman has it), or other causes — in my case, I have no idea.

I went to a semi-medical place for it once. I will not mention its name, for reasons which will become clear.

Imagine a classroom full of people on folding chairs, all waiting to be seen by the "doctor". On the walls were gory pictures of wounded feet and legs, and the words PERIPHERAL NEUROPATHY. People with PN can become numb to pain, and not realize they have a scrape or sore; an untreated wound gets worse.

As I understood it, their treatment (which cost $4,000) was to take out some of my blood, spin it around in a device, and remove the fat cells. These would be manipulated and put back in my body, to supposedly deal with the PN. It sounded reasonable — but <u>was not FDA-approved</u>. That matters, a lot.

There were three women in Florida who had poor vision. They reportedly had a treatment very similar to the one I was seeking. Their blood was manipulated, fat cells removed and put back into their eyes. Instead of being "cured," they lost more vision, and are now essentially blind.[1]

When I asked to see the doctor, we were taken to another room. Presently a man in a white lab coat came in, twirling his stethoscope.

We had a conversation, not a very long one.

Me: Are you a doctor?
Him: I am a chiropractor. (He stopped twirling the stethoscope.)
Me: Is your method FDA-approved?
Him: No, not right at the moment.
Me: Then you are required by law to post a sign saying it is **not** FDA-approved.
Him: No, we are not,
Me: Actually, you are.
Him: Excuse me a moment, I will be right back.
(He left.)
Gloria: I don't think he is coming back.
Me: I don't think he is coming back.

He did not come back. After a while we got bored waiting, and we left too.

The FDA needs to be strengthened, so questionable places like that can be investigated, and, (with sufficient cause), be shut down.

---

[1] https://www.nbcnews.com/health/health-news/three-women-blinded-bogus-stem-cell-treatment-florida-n734091

**Cancer.** I had (may still have?) prostate cancer. To fight it, I had surgery, radiation treatments, and hormone injections. Is it gone? I hope so. I take blood tests every four months, and so far, so good. One doctor said I might die **with** the prostate cancer, but not because of it. Usually, prostate cancer is slow growing.)

**Stroke.** I had a minor stroke. It was very strange. When it happened, I would actually *see* the letters of the words I was thinking, like s-t-r-o-k-e, before I could say them. That part lasted only a couple days. But there were other problems.

My childhood stutter came back, "blocking". That means you open your mouth to speak, and nothing comes out. That has not gone away. It is awkward at meetings sometimes. But there are techniques I can use from high school speech class, like starting the airflow and then joining it, like "Aaaaaare you all right?"

Is it unusual to have several chronic conditions? Not really.

One statistic cites forty per cent of Americans have two or more chronic diseases.[2]

What about you?

Do you have any of the following: multiple sclerosis, Huntington's disease, paralysis, Parkinson's, ALS, heart disease, kidney disorders, diabetes, epidermolysis bullosa, hardening of the arteries, immune deficiencies, deafness, blindness, Alzheimer's, HIV/AIDS, lung disease, COVID-19, bone loss, healing difficulties, SMA, sclerosis of the liver, obesity, leukemia, premature aging, IPEX, autism, Zika, brain tumors, sickle cell anemia, Urinary Incontinence, epilepsy, scleroderma, depression, schizophrenia — or something else?

Even if you have none of the above, chronic disease still affects you, and everyone: not only we ourselves, and those we love, but total strangers who need government relief (whereby we pay taxes) — *we need weapons to fight back* — and CIRM is the best of weapons.

Let me show you something wonderful. Glance back at that list of diseases.

If you Google any of those conditions, along with the word "CIRM", you will find California scientists fighting to defeat it. (Or, just go to the CIRM website.)

---

[2]—https://www.cdc.gov/chronicdisease/about/index.htm-

Example: I Googled one of my conditions: "peripheral neuropathy" and the word "CIRM," and immediately found these scientists working for its cure.[3]

At the Cedars-Sinai Medical Center, Dr. Robert Baloh is working on peripheral neuropathy, caused by a defect in the genetic makeup.

"The (scientist) proposes to …correct the defect in these cells…(then) he will transplant the (changed cells) into a rat model… to determine whether the cells…will alter the progression of the disease."[4]

Dr. Baloh made stem cell lines of a skin sample from a person with PN. Those stem cells were altered to become Schwann cells, which make myelin, the nerve insulator. These Schwann cells were then injected into rats with a nerve disorder — like mine — and the rats were improved.

My own conditions are easily borne, inconveniences, nothing more. Other people suffer terribly; I don't know how they get through the day.

CIRM scientists are working to cure what has been incurable until now. And not just the one condition, either! More than 80 therapies are already in CIRM-supported clinical trials. Also, there are another two dozen or so that we funded in early research, which then went on to clinical trials.[5]

CIRM directly funded portions of those 70 trials. In roughly 30 other cases, CIRM helped the research earlier, adding up to more than one hundred trials.

Nothing is easy; nothing is quick. But CIRM-funded research is developing repeatable procedures, step by step, to be sure what works, and what does not. Papers are published; records are kept, knowledge is gained and shared. It is being done the right way. Even the failures are helpful, making clear what should not be done.

Proposition 14 will keep that program alive.

---

[3] https://www.cirm.ca.gov/our-progress/awards/human-ipsc-modeling-and-therapeutics-degenerative-peripheral-nerve-disease
[4] https://www.cirm.ca.gov/our-progress/people/robert-baloh
[5] https://www.cirm.ca.gov/clinical-trials

# Photos

Author feeding Sevengill shark

Don Reed and Yaka the orca

Author enjoying nearness of lion

Gloria at right, holding tail of nurse shark

Roman still has "guns"

On August 25, Roman Reed b

Roman was first person to drive wheelchair over rebuilt Golden Gate Bridge

Don, Gloria and stem cell advocate Nancy Pelosi

Still the soccer ace Terri Reed

# 34 Legacy of Cure

Evie Vaccaro, girl in a plastic bubble, but only playing — she no longer has to stay in a germ-restricted environment. Thanks to parental dedication and CIRM-funded science, Evie's repaired immune system can now protect her.

Imagine if you could somehow listen to the reading of your own will! (That would be somewhat difficult, I admit) and the executrix reads what you wanted to be said: "To my beloved_____ I leave..."

What will we leave our descendants?

From me, not a lot: when I am gone my heritable goods will be whatever portion of my house that's not mortgaged, a car that may or may not start, several thousand very-used books; that's about it.

But there is one thing you and I might be able to leave, which the richest person on earth does not currently own — the cures for chronic illness and injury.

Rosie Barrero was once legally blind. Today she can see shapes, colors, and her children's faces. Her therapy was developed by CIRM-funded scientist Henry Klassen, who with Jin Yang started a new company, Jcyte, Inc. (blog.cirm.ca.gov)

What an inheritance that would be! Whoever owns that cure would instantly become the world's most popular person — "Fix me, heal me, give me the cure!"

What would it mean if we passed Proposition 14, the California Stem Cell Research, Treatment and Cures Initiative of 2020?

That could be the greatest inheritance imaginable.

Look at the mechanics. Every research scientist needs funding, right? CIRM makes serious funding available.

Here's how it works:

1. CIRM puts out a call for projects on a certain area of study; this is called an RFA — a "Request for Applications".
2. Scientists offer their ideas;
3. These projects are evaluated by a panel of out-of-state reviewers;
4. Reviewer recommendations (for or against) are passed on to CIRM scientists, who also review it. But they do not decide its fate; their opinions carry weight, but they are not the deciders.
5. The projects are put before the CIRM Board of Directors (technically the Independent Citizens Oversight Committee, ICOC) who say, "Go" or "No Go" — after public input.
6. If approved, the money will be given to the scientist — in limited instalments — as long as the milestones of progress are met, step by step. If the scientist cannot do what he/she agreed to do, the money for the project is cut off, all unspent funds to be returned.

7.  But if a therapy convinces the FDA, and the FDA grants the coveted Investigational New Drug (IND) status? Tests and more tests, after which, if all goes well in the clinical (human) trials, the scientist can either set up a company to develop the therapy, or sell it to a biotech enterprise.

What it boils down to is simple, and vital: CIRM provides funds for research and tests; without funding, nothing happens.

As mentioned earlier, I had scheduled an appointment to give blood. So yesterday I went to the Red Cross building, to do the deed.

First, I had to fill out a form. It was a lot of questions about my medical history — did I ever have Bubonic Plague, had I recently vacationed on Mars — to be sure my blood was healthy.

All went well, until the nurse took my blood pressure. She said: "Your heart skipped a beat five times in one minute...

My blood was fine, no problems there, but was my heart okay?

I called my medical provider, as soon as I got home. Several hours later, a doctor did a phone interview with me. He said there was probably nothing wrong, but I should have an electrocardiogram (EKG) to see if my heart was okay. (I did, and the results came back fine; I am healthy as a horse.)

Here is my point. What if there was something seriously wrong with my heart, but:

a. there was no cure;

OR —

b. What if CIRM had developed a repair for the damaged or diseased heart?

We know there has been progress. For example, a therapy has been developed to fight Duchenne muscular dystrophy-related heart disease: read the next two sentences:

"Capricor is using donor cells derived from heart stem cells developed by Cedars-Sinai to treat patients developing heart failure after a heart attack.

"In early studies the cells appear to reduce scar tissue, promote blood vessel growth and improve heart function."[1]

---

[1] https://www.cirm.ca.gov/our-progress/disease-information/heart-disease-fact-sheet

Will that particular approach be the answer? I have no idea.

But if I was indeed diagnosed with a heart problem, I would feel a whole lot better if the doctor said, "Oh sure, we know what to do with this!", instead of, "Sorry, pal, nothing we can do!"

The difference, of course, is research for cure. And if some usable therapy could be found? What an amazing inheritance that would be, from California scientists to the world!

A legacy of cure...

# 35 A Teenage Gorilla — And the Citizen's Initiative?

The gorilla's habitat must be protected, or this peaceful species will soon become extinct. But for sheer physical power, imagine pitting human strength against a gorilla's? (Gorilla Diorama, Atlas Obscura)

A friend of mine once fought a gorilla.

Personally, I tend to be diplomatic around creatures with many times my strength.

In my Marine World days I once had the joy of playing with a baby gorilla, by which I mean I sat on the ground and she climbed all over me, as if I were a tree. The gorilla cub was about the size of a full-grown chimpanzee, but thicker, and the muscles where they brushed against me

American champion and three-time Olympian Fred Lowe, one of Earth's strongest men. (Photo from *Strength and Health Magazine*)

were dense, and solid. She poked through my hair, apparently searching for fleas, then sat down abruptly on my lap, and stretched.

"Look at the inside of your wrist," said the gorilla's handler, pointing to thin white "wires" which showed when I closed my hand, "Your ligaments, right? Now compare that to hers."

My ligaments were thin as a pencil lead; hers were thick as my thumb.

"That's where the power is," said the handler, "In the ligaments."

Presently the gorilla got bored, swung up on the handler's back in a blur of motion, and they continued their walk.

But that is not the story.

Another time, scientists were trying to measure the strength of a full-grown male silverback. They loaded up a heavy-duty barbell with several hundred pounds and several keepers shufflingly carried it into the cage. The gorilla ignored it. But when an attractive female was relocated in the pen next door, and the barbell was leaned against the connecting open door, the gorilla looked at them, like you just had to get your way, didn't you? He put one hand on the bar — and tossed it.

But that is not the story either.

I will get to it in a minute, I promise, so stick around; it's a good one.

But first, what is the point of all this gorilla talk?

Imagine you are going into a big barroom brawl, and you could, if you chose, have a full-grown gorilla fighting on your side. That would be desirable, yes?

Now: imagine you had a terrible disease, like cancer, Alzheimers, blindness, deafness — as do millions across our country and around the world.

But you could, if you wanted, have the medical equivalent of a whole forest full of gorillas on your side — doctors and scientists, technicians and professors, intellectually the strongest people in the world — dedicated to finding a cure for whatever your condition...

People have that power right now, but it may be taken away.

It is called a "Citizen's Initiative", and it can be designed any way you want.

My favorite, of course, is Prop 71, the ballot initiative that created the California Institute for Regenerative Medicine (www.CIRM.ca.gov) but that is not the only initiative.

The Center for Disease Control and Prevention (CDC) lists a bunch of health-related initiatives, some of which might interest you.[1]

Proposition 10, the brain child of TV star Rob Reiner, put a 50 cent per pack tax on cigarettes, the receipts to improve the health of children, newborns till five.[2]

BTW, one key player in that effort was James Harrison, long-time attorney for CIRM and the ballot initiative campaign that created it, and credited by Bob Klein with helping to write our stem cell program.

There is also Texas, with Lance Armstrong's wonderful cancer-fighting program (CPRIT), also with three billion dollars.[3]

Unfortunately, some would deny the right of Americans to have such initiatives. And how would that happen? Just pass a law to take away our rights.

And that is what one political party is trying to do.

"So far (this year), Republicans have introduced 144 *bills to restrict the ballot initiative processes* in 32 states."[4]

---

[1] https://www.cdc.gov/publichealthgateway/strategy/index.html
[2] https://en.wikipedia.org/wiki/1998_California_Proposition_10
[3] https://www.cprit.state.tx.us/grants-funded
[4] https://www.nytimes.com/2021/05/22/us/politics/republican-ballot-initiatives-democrats.html

It is a safe bet that every patient advocate — for every disease — would be glad if there was more research money.

Maybe the initiative process could be a benefit for your state.

But if you want the initiative process to continue, **vote carefully**: because one party intends to destroy the citizen's initiative, and many other voting rights; the other party supports those rights. (More on this later).

And now, as promised, the gorilla story.

A bunch of the strongest people in the world (Olympic-style weightlifters) were visiting in the home of Bob Hise, then the leader of California lifting.

Among those present was the great champion Fred Lowe, who at 165 pounds would one day lift an American Record 402 pounds overhead. He was short, massive, and rock-solid.

Also present was (let's call him) "Chicago Friend", (CF) — and he said: "Freddy, rotate your shoulders inward, and crouch down".

Fred Lowe went along with the gag, for a moment, and the posture made him look even more huge, and slightly frightening.

"That's what the gorilla looked like," said CF.

It was at a county Fair, and in those un-regulated days almost anything could happen, if there was money involved.

A trainer had a ring with a padded mat, and for a fee you could *wrestle his gorilla*.

Now CF was a power lifter, meaning he did lots of squats, bench presses and deadlifts — he was extremely powerful.

But he was also young, and his judgement was perhaps not fully developed.

He noticed how the gorilla would do the same thing every time, grabbing his opponent, pinning him easily to the mat. He would let the human struggle for a few seconds, and then the trainer would say "Release!" Then the ape would let go, and take his pay in peanuts.

Nobody had a chance in wrestling a partly-grown "teenage" gorilla.

CF went into the pen — and *punched* the gorilla.

"The last thing I remember," CF said, "Was that his head did not move on his shoulders at all. It felt like punching a sack of sand. And he just looked at me."

When he regained consciousness, CF's friends told him how the gorilla had grabbed him by the ankles and slung him around, slamming him to the mat like a rag doll, and the trainer kept shouting, "RELEASE, RELEASE, RELEASE…"

The right to vote, and the citizens' initiative — these are powers on our side, like a friendly gorilla — and we must not let them be taken away.

# 36 Protect the Nerves!

Erika Miner has multiple sclerosis; her mother Karen Miner has a spinal cord injury. Fighting for our loved ones is what incites advocates.

One of my favorite aspects of California's hoped-for renewal program is a "set-aside" of $1.5 billion dollars for central nervous system (CNS) diseases.

Bob wrote this unusual inclusion into the program based on several conversations he had with scientists, doctors and others on the "listening tour" he and the Americans for Cures team did to develop the content of Prop 14.; Bob felt cancer therapies were moving forward strongly toward success; he wanted to be sure the complicated conditions of brain, spine and nerves were not neglected.

Everyone suffering a neurological disease has a reason to cheer that decision.

Day before yesterday, I called my friend Karen Miner about something, and she said:

"Well, you know Erika has Multiple Sclerosis (MS)."

It was like a boulder dropped into a pond.

I didn't add much to the conversation after that: just asked a few questions, and added MS to my list of diseases to hate.

Multiple Sclerosis (MS) is like a slowly developing spinal cord injury. For every individual, the symptoms are different. For most, when it begins, you are just tired for no reason, and it's harder to walk.

It is usually progressive. Things get worse, even shifting into another form of the disease, until paralysis sets in. And there can be pain.

If somebody asks you, why did Proposition 14 set aside one and a half billion dollars for CNS conditions, you could answer with names like Huntington's, or Alzheimer's, or Stroke or Spinal Cord Injury, and more.

Or you could say, it's for people like Erika.

It is said that soldiers in war do not fight for noble principles like love of country; they fight to protect their fellow soldiers, the person beside them. And so it is with patient advocates.

We fight to protect people we love. They are why the battle goes on. Before Karen told me about Erika, MS was not quite real to me, just words on a list of chronic diseases. Now, it is a threat to someone I care about: someone who has freckles and vivid blue-green eyes and a great laugh, and I went to her wedding.

As to the fight:

Nerves must be protected by a natural insulation called myelin. If the myelin sheath is not wrapped properly around the nerve, diseases like MS can occur.

Listen to an expert: Peter Schultz of Scripps Institute.

"MS is a painful, nerve-degenerative disease... the myelin sheath that insulates neurons is destroyed, resulting in loss of proper (nerve) function.

"Patients are often forced to stop working because their condition becomes so limiting...(interfering) with the ability to perform even simple routine activities, resulting in decreased quality of life.

"Existing treatments for MS delay the disease progression and minimize symptoms.

"However, the disease invariably progresses...

"We are pursuing an alternative therapy (of) regeneration of the myelin sheath … . The goal is to identify novel (new) myelinating drugs... Such drugs would be used... to prevent disease progression and restore proper nerve activity.

"(This would) represent a major unmet medical need; it could impact the roughly 50,000 Californians suffering from this disease, and reduce (the negative) economic impact on the state."[1]

And that does not count the 49 other states in America, nor all the other countries in the world.

These are not faceless statistics, but members of your family and mine — our loved ones. They are why we fight.

P.S. The following is from Erika: like all patent advocates, she is an expert on her own condition:

"I am 34 years old, married, living in Northern California, and serve as the COO for a nonprofit focused on providing equitable education to all students in California.

"I began experiencing a strange buzzing sensation in my left buttocks starting in September 2019. In December... I began experiencing an electric shock sensation down the back of my neck when I bent my head forward.

"This is called Lhermitte's sensation and is a classic symptom of MS. I made an appointment with my doctor, who referred me to a neurologist.

"My initial MRI indicated two spots on my brain that might be related to MS. To confirm, a plethora of blood work was ordered as well as a spinal tap. The spinal tap showed evidence of 8 oligoclonal bands. (A protein in the Cerebrospinal fluid — DR) The presence of these bands occurs in 95% of people with MS.[2]

With the presence of the "O" bands coupled with my MRI and negative blood results for viral or bacterial infections, I was given the definitive diagnosis of MS. By the time I received the diagnosis, nearly all of my symptoms had subsided.

---

[1] https://www.cirm.ca.gov/our-progress/awards/targeting-stem-cells-enhance-remyelination-treatment-multiple-sclerosis
[2] https://pubmed.ncbi.nlm.nih.gov/16945427/

"After extensive personal research and conversations with my neurologist, I began taking Copaxone, a 3x/week self-injection, to hopefully prevent the disease from progressing further."

(NOTE: I am neither scientist nor doctor, and am in no way qualified to recommend medication. However, Copaxone[3,4] has gone through clinical trials. One study showed a 34% reduction in symptoms. It is FDA-approved. DCR)

"As of today, in July 2021, I have been living with MS for over a year and been on treatment for nine months. I'm happy to report that nearly every day has been symptom free and I mostly forget that I have MS (the injections remind me every M/W/F though).

"No one can predict the future; but the future with MS is (even) more unpredictable. I hope that I am able to avoid flares in the future and that my symptoms never progress to something worse than the mild buzzing sensation that forced me to wonder if something was wrong."

— Erika

---

[3] https://multiplesclerosisnewstoday.com/copaxone-multiple-sclerosis/
[4] https://www.nationalmssociety.org/About-the-Society/News/FDA-Approves-Two-New-Generic-Forms-of-Copaxone%C2%AE-(G

# 37 Of Ninja Women Warriors, and The Fight Against Disease

Kunoichi, the female ninja. (NHK World-Japan, www3.nhk.or.jp)

John Cashman seeks to defeat pancreatic cancer. (www.chemregen.com)

Jionglia Cheng says: "Pancreatic cancer [...] soon will be the second most common cause of mortality due to cancer." — Kevin McCormack, CIRM.

Due to their secretive nature, (and the advantages of a fearsome reputation to terrify one's enemies), there is not a great deal of verifiable background about the ninja, shadow warriors of ancient Japan.

There is some fascinating information, however, about a league of women ninja, *kunoichi*, led by one Mochizuki Chiyome,[1] who recruited her warriors from among the economically disadvantaged: widows, orphans, prostitutes, women alone. She gave them work, training, a career, independence, and a sense that they were not alone.

The kunoichi worked as spies, information-gatherers, messengers, blackmailers, and, occasionally, assassins.

One of their most subtle maneuvers was called "the grass", in which beautiful women would infiltrate an enemy village. Naturally, they would be approached by unattached village males, and would soon be married, or become servants in the house of a leader, whom they would spy upon.

Attracting little notice, quiet as the grass, they might wait years, even bear children, but at a certain point a signal would be made — and the Ninja women would spring into action, killing the men they had moved in with — and whomever else was on their list.

---

[1] https://en.wikipedia.org/wiki/Mochizuki_Chiyome

Emotionally, I would be on the Ninja women's side, because of the generally rotten treatment women have received throughout history. But if someone was attacking my family? I would have no choice but to fight back.

Now. As you know, I have a grudge against *pancreatic cancer*, that miserable disease which took Gloria's life. As long as I live, I will be alert for those two words, like when a guard dog's ears twitch, if it hears an important sound.

That's how I felt reading Kevin McCormack's excellent short piece[2] on John Cashman's and Jionglia Cheng's research. Kevin is communications director for CIRM, and the kind of writer one hopes for — if you see his name on something, read it; you will invariably find it quick, understandable — and it will matter.

Working on a CIRM grant, Cashman and Cheng have an ambitious goal: to eradicate cancer, all kinds including: pancreatic, prostate, colorectal, and breast.

The Cashman-Cheng weapon is PAWI-2, which stands for: p53 Activator WNT Inhibitor-2. It was developed after doing medicinal chemistry on hundreds of similar molecules.

"PAWI-2 is a small molecule, like a drug, and contains a sulfonamide group. As a class, sulfonamide drugs in use today are non-toxic and, thus far, PAWI-2 shows great safety in small animal studies." — Dr. Cashman, personal communication.

If successful, PAWI-2 will *fight cancer five ways*. First, it will discourage the cancer growth. Second, it will actively attack cancer *stem* cells. Third, it will challenge the "pathways" used by cancers as they multiply. Fourth, it will encourage apoptosis, natural cell death, of the cancers. Fifth, it will "synergize" (work together with) other anti-cancer therapies in use today.

Overall, their invention seeks to wipe out the cancer: not to lessen it, but to erase it altogether, so the body is clean of the microscopic killers.

The target of PAWI-2 is cancer stem cells (CSCs), which, like the deadly Ninja women, are hard to find, and difficult to kill. The CSCs can survive attacks of chemotherapy which kill lots (but not enough) of the "normal" or bulk cancer cells. If the cancer stem cells are not killed, they will soon make more of the others.

---

[2] https://blog.cirm.ca.gov/tag/jionglia-cheng/

Only rat and mouse studies have been done: but already "...PAWI-2 demonstrated effectiveness in blocking breast, prostate, and colon cancer. When tested in the laboratory, PAWI-2 showed it was able to kill pancreatic cancer stem cells, (including some which were) drug-resistant."

One of the anti-cancer drugs used by my wife was Gemcitabine. Some scientists call it "ineffective", and I must agree. But PAWI-2 might synergize with Gencitabine and make it very effective.

In one of their major papers, "PAWI-2: A novel inhibitor for eradication of cancer", the authors state:

"...PAWI-2...causes...cell cycle arrest."

"...cell cycle arrest"? Stopping the cycle of cancer stem cell multiplication...killing the cancer —

Like the ninja women warriors, if they were fighting on our side!

---

[3] https://tinyurl.com/y5uawv88

# 38 Son of Bob Strikes Back

Like his father, Rob Klein III has his foot in two worlds: real estate and stem cell research. (https://www.worldstemcellsummit.com/robert-n-klein-iii/)

Bob Klein's strength is also his weakness. If you looked at notes of a meeting with Bob, you would see he generally ends up with a deskful of new chores. He volunteers much better than he delegates. And (while not as ancient as the author of this book), no one is as young as we once were.

We were approaching the climax of the Prop 14 battle. Bob was working like a one-man committee, as if there were several of him, never refusing a chore or a challenge.

But when Bob gets tired, as he was now, you can tell; his face changes color, turning visibly gray; Everybody told him, "Take it easy, Bob, don't work so hard!" — which is, of course, the one task he cannot do.

One day Bob called a Zoom meeting, very small, just six people: Bob, his son Robert, Paul Mandabach who worked with us since the beginning, Bernie Siegel, founder of the World Stem Cell Summit, James Harrison, Bob's friend and super-lawyer, and myself.

And at the meeting Bob said: "If anything should happen to me — like if I was out riding a horse, and — well, something happened — Robert takes over."

There was an utter stillness in the room. It was like Bob to face every problem, including his own potential demise. He had always been the irreplaceable man. But when his time ran out, as it will for us all?

The silence lengthened. It seemed I should say something, but what? I could only blurt out:

"I cannot imagine a world without Bob Klein. But if there has to be — Robert is eminently qualified; he is the right person; he will get the job done."

Like his father, Robert "Rob" Klein III has a foot in two worlds: real estate, and stem cell support. He is strongly built, cheerful, a face not complete without a smile. When I call him with a problem, which is fairly often, he will say "hmm", and then, "Well, let's take a look at that." If he can't solve it on the spot, he won't pretend he has an answer. He will just say "Let me get back to you.", and he will.

Below is an example of why I have confidence in him.

An individual had written systematic criticisms of Proposition 14, and the entire structure of the program. I see no need to mention his name; he has a right to his opinion. It was a serious assault — but Robert responded. He wrote a detailed, lengthy piece answering the opposition's attacks.

I edited the response lightly, but the article is his throughout. It is somewhat lengthy, even cut down from its original 20 pages: a useful reference document.

"**CORRECTING THE RECORD**: The Opposition's Inaccurate Claims".

"The fundamental purpose of Proposition 14 is to develop early intervention therapies that mitigate disease, save lives, and reduce healthcare costs for every Californian and the State of California itself. The funding is provided through bonds that will be issued over a period of approximately 14 years...and be repaid over roughly 40 years, with payments not beginning until 2026.

"An opponent of Prop 14 has placed an editorial... in which were a number of factual inaccuracies and materially misleading statements about Proposition 14, and its predecessor Proposition 71.

"The information below is a fact check to correct the record.

**Opposition Claims:** "This research is already well funded." (implying there is no need for CIRM's funding — DR)

**The truth** is: "...NIH (National Institutes of Health) primarily funds discovery and translational research, and participates in very few clinical trials, (the expensive part) as well as no clinical trials to develop... embryonic stem cell treatments. Without CIRM, many California Research institutions would lose almost 50% of their stem cell research funding, and some losing more..."

**Opposition Claims:** "CIRM already is costing the state $327 million a year to repay the bonds authorized by voters in 2004, via Proposition 71."

"**The truth** is: The California Treasurer's Office report shows...Proposition 71 will cost an average of about $76 million a year (with a population of 40 million, about $2 per Californian annually) over the remaining payback period...clearly way under the $327 million number claimed by the opposition.

"As for the Prop 14 bonds, the average annual payment is projected to be less than five dollars ($5) per person, over the payment period."

**Opposition Claims:** "After spending all that money, not a single U.S. Federal and Drug Administration (FDA) approved product has materialized on which CIRM's funding played an important role."

**The truth** is: "This is fundamentally wrong, and misleading. In addition to two FDA-approved cancer treatments, Inrebic ® and Daurismo ®, CIRM has been credited with 2 breakthrough therapy designations, one for another cancer therapy, and the other for a previously fatal immune system disorder. In addition, CIRM funded 7 therapies which the FDA recognized as deserving of their RMAT status (Regenerative Medicine Advanced Therapy designation), which allows deserving medicines to move forward more quickly. These designations are highly difficult to acquire, and only a very small proportion of research projects receive them.

"More than 100 clinical trials have been funded by CIRM, many of which are already demonstrating high levels of efficacy, and which may also receive the coveted RMAT designation."

**Opposition Claims:** "That means California taxpayers will be on the hook for $587 million a year for stem cell research."

"**The truth** is: this figure is based on outdated and incorrect information.

"There are no bond payments at all for the first five years. An additional benefit is that for at least the first ten-to-twelve years, Prop 14 will generate more state tax revenue than the cost of the bonds, and like Prop 71, stimulate billions in economic activity. The economic benefits far outweigh the costs.

"The highest combined debt service will be the 11th year, $221 million (a little over $5 per Californian), after which Prop 71 will be almost entirely paid off.

"A surprise to the critics is the bond allocation limits. Bond sales for stem cells cannot exceed one half of one per cent, (0.5%) of California's total bond sales in a year, leaving ninety-nine and a half per cent available for any other purposes the state may consider a needful priority."

**Opposition Claims:** "Nor has the state achieved any of the projected $3.4 to $6.9 billion in direct health care cost savings when the original measure was passed."

**"The truth** is: these savings were based on a *35-year projection*, with such benefits not expected before 2039. Additionally, this date must be pushed back due to the funds being frozen by litigation till 2007.

"A number of FDA priority-ranked (and CIRM-funded) trials are expected to result in FDA-approved treatments in the next few years. It would be irresponsible to abandon these promising curative treatments now, when California's therapies are so close to the finish line.

"Chronic disease is the leading cause of death and the leading driver of health care cost increases, and medical bankruptcies, Prop 14 is a small price to pay; potentially saving millions of lives and billions of dollars over the coming years.

"A study examined the cost of 8 chronic diseases and conditions — for which CIRM funding had led to ground-breaking therapies in trials — Type 1 diabetes, sickle cell disease, spinal cord injury and others. It was determined that cures for just these and improved therapies for others could bring a savings of about $47 billion over the next 30 years."

**Opposition Claims:** "CIRM was created by a ballot measure in 2004 because the federal government had restricted funding for human embryonic stem-cell research."

**"The truth** is: fighting back against restrictions on research was indeed part of the reason for Prop 71's appeal, but only a part: more important was to accelerate the development of therapies, treatments and cures, as well as making California the center of the stem-cell universe, and to thereby attract the world's leading scientists to advance stem-cell research and cures.

"Overall funding in stem cell research was too low, inconsistent, and unreliable.

"Threats against research freedom remain, and must be taken seriously, as in the recent letter from 22 ideologically driven extremely conservative U.S. Senators to (now former) President Trump, asking for a ban on embryonic stem cell research. "Also, it must be acknowledged that the federal government has funded exactly zero clinical trials for embryonic stem-cells in California or any other state.

"There is a huge gap in funding. The NIH will fund basic research, and private enterprise will get involved after proof of efficacy has been established — but funding for those all-important clinical trials...? Almost nothing.

"California is the only state with the scale, infrastructure, and scientific resources to broadly advance stem-cell clinical trials, and meet that shortfall in funding, which kills many medical research projects, and is called the "valley of death".

"*Ranked alongside nations*, California is number 2 in the world in biomedical research capacity. Only America (including California!) exceeds CIRM."

**Opposition Claims**: "Nor does it address the widely criticized conflicts of interest among board members, the majority of them representing institutions that have received the bulk of CIRM's spending."

**The truth** is: "This statement is baseless. In 2012, the already strict rules were enhanced, so that *no one who represented a research institution would be allowed to vote on any grant*. The opposition also does not mention the out-of-state scientific review board which will study, analyze and comment on every grant proposal — and recommend for or against funding — to reduce the chance an evaluation may be done by someone with a personal connection to the scientist.

"CIRM is also the ONLY state agency with an independent oversight committee, the Citizens Financial Accountability Oversight Committee (CFAOC) which is chaired by the state Controller. CIRM also undergoes financial and performance audits by the Controller's office and the Bureau of State Audits.

The primary source of the opposition's conflict of interest claims is from a study produced just a few years into CIRM's existence. The study (commissioned by CIRM itself) was intended to strengthen the Institute's conflict provisions even further. The suggestions were highly hypothetical — yet nearly every one was adopted by the CIRM board within a year."

**Opposition Claims**: "It does nothing about the absurd requirement that any changes to the agency require a 70% super majority vote of the legislature."

**The Truth** is: "The 70% vote requirement is not unusual; of the first 20 bills in the 2019–2020 session, 17 received more than 70% of votes in support. This simply insures that changes will not be made lightly, on … political whims.

"Such legislation is not only possible, but some has in fact been passed, and with the full cooperation and participation of CIRM staff, leadership and relevant experts. Nearly all involved agree that changes within this legislation strengthened the Institute's ability to best support the research and fulfill the voters' will.

**Opposition Claims**: "In 2004, the backers of the campaign said they anticipated returns to the state of up to $1.1 billion; to date, CIRM has returned $350,000."

**The Truth** is: "Again, the opposition is using old statistics. More up to date figures reveal that CIRM-funded research has generated at least $4,612,433 as revenue for the state, more than ten times the amount of the figure stated (and has grown still more since this document was made.).

"The real source of financial benefit for the state, however, will be reduction of medical costs through early intervention, mitigation of suffering, and cure. There will also be new revenues from businesses brought about by the economic stimulus. But above all, the diminution of suffering and medical incapacitation will have a financial impact — fewer attendants, less time in hospital, fewer hours lost from work, and a smaller number of people forced to live on government assistance. And today? Nearly 900 intellectual property filings have been either granted or are pending — leading to new businesses and greater state and local revenue-generating economic activity for California.

"Since the campaign ended, a number of additional partnerships, acquisitions and new ventures have been announced and/or have gone through, This figure has grown literally to billions, with much more expected by the end of the year (2021). As this will be an exponential curve, far more is expected in the years to come."

**Opposition Claims:** "It would require that any returns from the state's investment in new therapies are given back to pharmaceutical and biotech companies, thus freeing them from any price restraints, with CIRM making up the difference."

**The Truth** is: "This is an egregious manipulation of the facts. No such requirement is made, anywhere in the text of Proposition 71 or 14. On the contrary, before any new CIRM-funded products are released to the market, their inventors must design a low-income affordability plan, to make the product available to those who might not be able to afford it. This plan must be approved by the Independent Citizens Oversight Committee (ICOC) the CIRM's governing board.

"Indeed, an unparalleled provision: a dedicated staff will work to enhance affordability and access to procedures and therapies developed by CIRM. This committee will work to ensure that insurance companies, public or private, and California's governmental program, Covered California, as well as other major payers, will provide early coverage for patients and doctors seeking to utilize the new therapies.

"Patients wishing to participate in CIRM-funded clinical trials, but who cannot afford the expenses involved (the therapies themselves are free) will find their way made easier; help will be offered, including assistance to cover expenses for travel, meals, caregiver costs, lodging, etc., making clinical trial participation possible for many Californians who otherwise could not afford (it).

Revenues from intellectual property developed with CIRM funding… will be used to make therapies more affordable. To characterize this as a corporate subsidy is disingenuous.

**Opposition Claims:** "If California is going to spend billions to fund stem-cell research, the legislature should draft a new measure that does it the right way.

**The Truth** is: "The "right way" is of course a matter of opinion. CIRM is based on the belief that science and medicine should drive medical discovery and therapy development, not a political agenda.

"These medical discoveries require a long-term process to move from theory to therapy; 12–15 years may be needed to move from discovery to application. Many projects would be abandoned, or not started in the first place, if there was no confidence in continued support of project-long funding. Matching funds would not be attained at nearly the same levels without reasonable expectation of reliable funding.

"If elected officials were allowed to start, stop or redirect funds every year or two, whenever they (wished), the impact could be devastating.

"Scientists need to have a reliable source of funding, if they are to do their work successfully. Denied that, many might choose to go elsewhere in search of more reliable funding. Some already have…"

**Opposition Claims**: "The returns are limited because by law the state cannot hold equity and act like a venture capitalist would do. Amending that rule should have been included in the new measure".

**The Truth** is: "Proposition 14 is an initiative, not a constitutional amendment, which would be required to implement such a suggestion. To turn the government into a venture capitalist is a highly complicated and controversial change. It would be outside the focus and objective of Proposition 14, which aims to develop treatments and cures for patients who need them.

"Instead of such radical changes, which would have threatened the passage of the initiative, the agency has developed constitutionally acceptable methods to increase the state's financial benefit. These include using options and/or warrants for stock and participating loan agreements, to increase returns to the state in case of a major financial success, which are then directed to help Californians afford the therapies that resulted from the research they funded."

Materials produced by Robert N. Klein III

# 39 The Villain in Aging?

I almost broke my hand on a helmet like this one. (brassbinnacle.com)

I live in a two-story house.

When I come downstairs in the morning, it is eight steps down, turn left, eight steps more to the living room. But right at that first turn is an old-fashioned deep-sea diver's helmet, the famous Mark 5, on a little stand. If you saw "BENEATH THE TWELVE MILE REEF", or any of the old diving movies, this is the helmet they wore.(Also Charlie Sheen has one in his living room in the TV show "Two and a Half Men".) I wore one in commercial diving school many years ago.

Stanford's Dr. Helen Blau fights aging with the science of muscle growth. (med.stanford. edu)

The other morning I could not find the light-switch. It was pitch black. But I was sure I could find the way, even in darkness. I took one step, and tumbled, spectacularly. Remember the helmet? Somehow, I managed to find it, with the *back* of my right hand. I don't know how. The pain was amazing. My knuckles felt broken, and I swiftly exhausted my vocabulary of expletives.

Reconstructing the accident afterwards, I realized my right leg had given out on me. I extended one leg, transferred my weight — the knee buckled, and I fell.

I added some leg-strengthening exercises to my morning routine — and am hunting for the light-switch a little more carefully now.

But I remembered something: *almost the same exact incident happened*, a couple years ago, but that time I caught myself. I just crouched a little deeper on one leg, and did not fall. I was just stronger then. Might there be a way to recover some of that strength?

I have known Dr. Helen Blau of Stanford for several years, and until now, it always seemed she was attempting the impossible.

Her goal was to ease the devastation of old age: to recover strength, muscle mass, and endurance.

I am 75, and physically on a downward trend. According to Dr. Blau, a typical human past 50 will lose 10–15% of their strength every decade — like your power diminishes one per cent every year.

But must we lose all that strength?

Not if a molecule called PGE2 turns out to be helpful.

PGE2 — remember it like Part Two of the electrical company — Pacific Gas and Electric II — (no actual connection, just the memory device) PGE2.

When Dr. Blau gave PGE2 to old mice, it restored strength, muscle mass and endurance to their aged limbs. On the treadmill, they ran faster and longer.

But if the "hero" of aging is PGE2 (prostaglandin), there is also a "villain", a protein called 15 PDGH (hydroxyprostaglandin dehydrogenase).[1]

PGE2 makes muscles bigger and stronger; 15 PDGH withers the limbs. It does so by weakening PGE2, "*inhibiting*" the good stuff. So what do we do, to diminish the negative effects of 15 PDGH? We inhibit the inhibitor!

Now. Consider the misery old age can bring; would you mind terribly if we skipped some of that suffering, kept our strength longer, and grew older more gracefully?

Presently, Dr. Blau is working on a number of muscle-related projects. One uses PGE2 to strengthen the diaphragm, the muscle your body uses to breathe. This could help COVID-19 survivors regain muscle strength, which they lost from being on a ventilator.

There is also sarcopenia, an overall weakening of the muscles. This afflicts about 15% of the population, costing America more than $40 billion dollars a year in health care.[2]

And what about the simple ability of deciding when we pee? Renewing muscles might help an aged urinary sphincter recover control, as if it were young.

Everything we do depends on muscle — if it wastes away, so do we.

Can the substance be safe for human use?

"PGE2 is already being used to induce labor (childbirth) and to treat respiratory distress in newborns," says Blau.[3]

---

[1] "Local delivery of rejuvenated Old Muscle Stem Cells to Increase Strength in Aged Patients", CIRM Grant Number TR3-05501

[2] Sarcopenia, annual cost: https://pubmed.ncbi.nlm.nih.gov/30997923/

[3] "Small molecule restores muscle strength, boosts endurance in old mice", by Krista Conger, "Stanford Report" December 10, 2020.

Human clinical trials on new uses for PGE2 may start in 2022. If I fit the requirements, I would love to be part of those trials — and maybe gain a few more years of youth and strength.

During a recent conversation with Dr. Blau, she brought up new information.

Q: In your work with COVID, is the primary potential benefit a repair of weakness in the muscle wall of the diaphragm?

A: Yes, but there is more to it than that. When a person comes off a ventilator, the muscles have grown weaker (from inactivity). Depending on how long he/she was on the ventilator, there is a possibility the diaphragm will simply collapse. This can be fatal.

It's a big problem with COVID, because some people have had to be ventilated for weeks, causing the muscles to weaken substantially. We hope that we can use PGE2 to strengthen the diaphragm to mitigate that weakness. This would also make more ventilators available — by decreasing how long people are kept on them.

Q: Might there be a use for it in the battle against Spinal Muscular Atrophy (SMA), that deadly killer of children?

A: Yes. Any condition with muscle wasting and loss of strength might benefit.

Q: Might your therapy benefit people with Sarcopenia — overall weakness?

A: Yes. One very real fear is that the patient becomes so weak with Sarcopenia that he/she can't get out of a chair properly and may fall and break bones.

Our goal is two pronged — localized muscle strengthening, in specific areas, as to combat urinary incontinence...and global muscle strengthening: to increase the size and strength of muscles throughout the entire body.

Q: Would it be correct to say that your overall goal is to (help) the function of (withered) muscle — restoring power with increased muscle mass?

A: Yes. For example, Duchenne muscular dystrophy patients (have) severe muscle wasting. We might mitigate some of that loss of skeletal muscle strength. We do not yet know the effects of PGE2 on the heart; but it is heart muscle failure that makes patients die. PGE2 might restore the heart's strength (and function — DR.)

Q: Are the drugs FDA approved, which could save years of delay?

A: One is a novel drug and the other is a novel drug combination. So we still need clinical trials to get FDA approval and ensure that the drugs are safe and effective.

Q: Would there be military applications? Dr. Anthony Atala of Wake Forest is in charge of AFIRM, (Armed Forces Institute for Regenerative Medicine) and I will be talking to him soon.

A: Yes, absolutely. Wounded soldiers are often bedridden for extended periods of time. It is vital that they recover strength and mobility as soon as possible, to lessen the possibility of permanent damage.

Q: You have a spinoff company: Myoforte Therapeutics; what does that mean?

A: "Myo" is a prefix that indicates muscle (Greek origin), and "forte" is from the Latin and signifies "strength"

May Dr. Blau's work help people of a certain age maintain strength and independence — and decrease my likelihood of falling down stairs!

# 40 133 Million Reasons for Cure — and One More

Marwa Azzad not only fights spinal muscular atrophy herself, but as a pharmacist offers help and guidance to others with the brutal condition. Here she is in front of the famous Alsayed Albadani mosque.

Someone you know has a chronic disease: a condition defined as long-lasting, (at least one year) or incurable. Maybe it is you, and it is definitely me.[1]

Multiply that by approximately 133 million Americans...more than 40% of the total population of this country..."[2]

---

[1] https://www.cdc.gov/chronicdisease/about/index.htm
[2] https://nationalhealthcouncil.org/wpcontent/uploads/2019/12/AboutChronic Disease.pdf

Many have more than one chronic disease — and that is in America alone!

Now think of just one more: a child with Spinal Muscular Atrophy (SMA). It is a terrible condition, diagnosed at about six months of age, after which they become weaker and weaker, gradually becoming paralyzed.

I have known two such children: Gwendolyn and Pranav. Each required mechanical assistance to breathe, and a tube in their stomachs for food.

Both were beautiful, and beloved; their parents worked like Hercules, trying to protect their children.

Pranav's mother gave up her career to stay home and care for her son, full-time.

Gwendolyn's parents set up a foundation, the Gwendolyn Strong Foundation, to battle the condition.

Yet despite their heroic efforts, both children's lives were drastically shortened.

Gwendolyn and Pranav: hold them in your mind.

Now. Leap halfway around the world, to a child care center in Egypt, specifically Tanta Section 2 Child Care Center; there a young woman, Marwa Azzam, works as a part-time governmental pharmacist.

Marwa takes special interest in children suffering SMA. Doctors with SMA patients frequently refer them to her, so she can help in their daily lives.

"When I fill their prescription, I also give them advice," she says, "So they will not make serious mistakes on day-to-day living, diet and exercise."

Marwa, an adult, also has Spinal Muscular Atrophy.

I did not realize there are several different types of SMA, and that some allow longer lifespans. Marwa's email sent me back to the CIRM website.

I came across a CIRM document about the work of Clive Svendsen of Cedars-Sinai Medical Center. It said:

There is "a gene (SMN2) which can lessen the severity of the disease process. Children with more of this modifying gene can live longer periods of time..."[3]

---

[3] — https:www.cirm.ca.gov/public-web-disease-focus/spinal-muscular-atrophy

"To lessen the severity of the disease... (and) live longer periods of time..." — think what would have meant to the families of Gwendolyn and Pranav!

Marwa's life is not easy. She suffers from the disease, with its muscle weakness and frequent exhaustion, requiring a wheelchair.

But she is alive, and a useful member of the community.

"I live alone since my father died. I always hire someone, a Personal Care Administrator (PCA) to help with personal matters, just as disabled people do in your country. She helps me at home and outdoors.

"At work, my PCA takes the prescription from the patient and gives it to me. I read it, dispense the medication, and do patient counselling when needed.

"I love my work; it was my childhood dream to work in the medical field."

Marwa is especially interested in the possibilities of stem cell therapies. She feels they "may help regenerate atrophied muscles in SMA patients."

Remember that "good gene", SMN2?

Apparently, Marwa says, "I have ... more SMN2 copies, and therefore less severe symptoms...

"Also genetic drugs are more effective on patients having more SMN2 copies.

"...there are 3 available genetic drugs that compensate for the missing gene. But they do nothing (for) the atrophied muscles, so (we need) stem cells to regenerate muscles and boost the effect of genetic drugs." — Marwa Azzam, personal communication.

Could we develop more of that gene, SMN2, to strengthen its attack on Spinal Muscular Atrophy?

I am not a doctor, and do not know. But I definitely want scientists like Dr. Svendsen to keep the work moving forward — to hunt for cures for the 133 million Americans with a chronic illness or injury — and all the others, around the world.

For info on the multi-trillion dollar financial costs of American chronic disease, check with the Center for Disease Control and Prevention.[4]

---

[4] https://milkeninstitute.org/sites/default/files/reports-pdf/ChronicDiseases-HighRes-FINAL.pdf

But no one can put a number on pain. No financial statistic can sum up suffering, or the loss of a family member through chronic disease.

And globally? Around the world, one person in three suffers chronic disease.[5]

That's why the California stem cell program is fighting hard, so cures may spread, to meet the disease.

Because everyone has a right to be healthy — and a responsibility to help.

---

[5] https://www.ncbi.nlm.nih.gov/pmc/articles/PMC6214883/

# 41 "And care for him, who shall have born the battle"...

One of America's top scientists, Anthony Atala, was instrumental in developing the Armed Forces Institute for Regenerative Medicine. (school.wakehealth.edu)

As I drove to the VA hospital in Palo Alto, California, I thought of President Lincoln's words on America's responsibility for wounded soldiers:

"And care for him, who shall have borne the battle..."

He said it in the fewest possible words: simple, elegant, and true...

I was on my way to get my COVID-19 vaccination at the Palo Alto Veterans Administration.

Everyone I met at the VA was kind and thoughtful. When I got lost (a not unusual occurrence for me) everyone I asked for directions had time to stop and tell me. No one turned away; no one was too busy.

Soldiers like Lt. Col. Tammy Duckworth (ret.) deserve to have lost limbs restored. (Sen. Duckworth's website)

And when I got panicky because I could not find my Veterans card, they said, "Don't worry," and found it electronically. Small things, but they mattered.

A nurse said "Have a seat!", to wait for the injection.

I was getting my vaccination early because of my age (75) and veteran status.

I had volunteered for the Army in 1963, signing up for three years. The job they trained me for was to knock down enemy missiles by electronic counter-measures.

I was assigned to hot places: the Arizona desert (Ft. Huachuka), and a Texas swamp (Ft. Wolters). The desert was dry, the swamp miserable. As there were not a lot of missiles over Arizona or Texas, my chores were often make-work, to keep me and my platoon mates busy. I remember once digging a ditch, about 20 feet long and 4 feet deep. It was a very good ditch, I thought. But when we finished, the lieutenant said: "No, I changed my mind. I don't like it there. Let's see, move it over there." How do you move a ditch? Uh-huh. You dig it again, and again, until the people in charge think it is good enough…

When I had about a year to go, I was offered the chance to transfer to Viet Nam.

But by then I had seen enough to feel the people in power might not necessarily have my best interests at heart. I said no, and did the rest of my duty stateside. I saw no fighting, did no killing, for which I am eternally grateful.

There were others in the Palo Alto VA Hospital that day who were not so lucky. These were warriors, and some carried marks of their sacrifice.

One man came slowly but smiling, using his hands to push his wheelchair: his right leg was gone at the hip.

I thought of Illinois Senator Tammy Duckworth, who had lost both legs and the partial use of her right arm while flying a combat mission over Iraq, November 12, 2004. Lt. Col. Duckworth worked tirelessly as an advocate for her fellow soldiers, becoming Director of the Illinois Department of Veterans Affairs. In 2009, she was appointed Assistant Secretary of Veterans Affairs, and to this day she fights for the betterment of soldier's lives.

No one deserves cure more than she — and yet we have not made her well.

I thought of the lowly salamander, which can regenerate a lost leg, or a spine. Why can't we do what the salamander does, and regrow a soldier's missing limb?

The military, I am glad to say, has an organization investigating such ideas.

The Armed Forces Institute of Regenerative Medicine (AFIRM) is "dedicated to repairing battlefield injuries through... regenerative medicine technology"...[1]

Let me introduce one of the founders of that organization.

Dr. Anthony Atala has won stacks of awards, too many to list them all here. He was honored by Scientific American as one of the world's most influential people in biotechnology, and was similarly recognized by U.S, News and World Report. His work was listed (twice!) by Time Magazine as a top ten medical breakthrough of the year. He received the Edison science/medical award for innovation, and the Smithsonian Ingenuity Award for bioprinting tissue and organs.

Dr. Atala designed, built, and surgically implanted the world's first engineered organ – a bladder – whose owner is still using it!

He has published over 800 scientific articles, edited 25 books, and holds over 250 national or international patents. He is the Director of

---

[1] https://afirm.mil/assets/documents/annual_report_2011.pdf

the Wake Forest Institute for Regenerative Medicine, in Winston-Salem North Carolina. In short, he is one of the world's top scientists.[2]

In 2005, Dr. Atala spoke before the Combat Casualty Care Conference. He shared information about the state of regenerative medicine: things which might help wounded soldiers regrow tissue, or limbs.

The following year, Colonel Bob Vandre developed the idea of "a regenerative medicine institute similar to the Department of Defense... but aimed at near-term translational research...".

Paraphrased from their literature, here is AFIRM's challenge[3]:

1. Limb and Digit Salvage: to replace nerves, bones, tendons, ligaments and skin, <u>restoring arms and fingers</u>, blown off in battle;
2. Craniofacial Reconstruction: to <u>repair torn faces and broken skulls</u>;
3. Scarless Wound Healing: in addition to the emotional stress of disfigurement, <u>scars</u> can be so thick they restrict movement. We need ways to remove them, to make the skin clean and young and intact;
4. Burn Repair: to improve <u>skin substitutes</u> for burn wound grafting;
5. Genito-urinary and Lower Abdominal Injury Treatment: to rebuild <u>genitals</u>.

"Roll up your sleeve, please," said the nurse, filling up a syringe.

I did as I was told, and presently the needle slipped into my shoulder. It stung a little, but not like a harpoon or anything, just a regular little bee sting shot. I hung around 30 minutes, in case of an allergic reaction, but nothing happened, then or since, just a little soreness in the muscle.

And, so far at least, I did not get COVID-19.

I spoke with Dr. Atala recently. We chatted cheerfully, (it is not hard to get me talking about CIRM!) and then I asked him a question: did he have enough research money?

He hemmed and hawed a little, as if not wanting to hurt any funders' feelings, and then admitted there might be "some long-range projects" he had in mind...

That was what I wanted to know. Long-range projects are what CIRM does best.

---

[2] https://www.futuremedicine.com/doi/abs/10.2217/17460751.3.4.463
[3] https://www.afirm.mil/assets/documents/annual_report_2013.pdf

Right now, the Army is helping out with civilian COVID testing, sending military doctors and nurses to places of need.

I vote we return the favor.

Think of it: the California Institute for Regenerative Medicine (CIRM) and the Armed Forces Institute of Regenerative Medicine (AFIRM), combining our strengths — would this not be the ultimate expression of both groups' goals?

Let me say that again: CIRM and AFIRM should work together: peacetime folks and active military.

AFIRM could set up a medical research branch in California, (UCLA is already cooperating with the military to help wounded warriors) and apply for grants.[4]

What a benefit, for both civilians and the military — especially our country's wounded.

Do we not owe them the best treatments science can provide?

---

[4] https://www.afirm.mil/index.cfm?pageid=about_afirm.consortium.wake_forest.overview

# 42 An Appreciation of Asian-American Scientists

Making the Chinese-American play, *Legend of Wing Chun*, allowed students to learn about cultures other than their own.

Minorities seldom want to be singled out by their ethnic background, even for praise.

But the violence and persecution recently inflicted on Asian-Americans is too brutal to ignore. One television clip showed a big man knocking down a small elderly Asian woman, after which he kicked her *in the head*. It was a miracle she was not killed, nor paralyzed.

In previous books, I have written about racism endured by Latin-X and African-Americans. Similarly, I must say a few small things about the Asian-American community, from which so many top scientists have sprung, and what that group means to me personally.

As a teacher, I was delighted by the support I received from Chinese-American families. For example, I led a multi-cultural student club, where the kids gave up their lunchtime and worked all year on a play with minority themes (African-American, Hispanic, Asian-American), fun with a purpose. I wrote and directed; the students performed; proceeds went to Christopher Reeve.

When word got out that our next play would be on Chinese-American history, I was amazed by the response.

A local Chinese-American club asked me to speak at one of their dinners, at which they gave the play a financial donation ($600), and pledged their support. These were not empty words.

For instance, we were trying to make a costume for the Dowager Empress, Cixi Taihou. The community rented an authentic replica from a museum. On the night of the performance, parents came and did the actors' makeup. One parent took professional level pictures, and made them into a photobook for me, which I treasure.

At the end of the play, the entire cast and crew came on stage. And while the famous martial arts music "General's Mandate" played, we all performed the first empty-hand set of Wing Chun: Siu Nim Tao, which means "small understanding".

The play focused on the evils of racism — like the Chinese Exclusion Act of 1882. Until former President Trump tried to ban all Muslims from entering America, the Chinese were the only ethnic group to be systematically blocked from entry. The law was imposed on the grounds that Chinese workers were taking jobs away from white men — although Chinese were only .002% of the population. Chinese were not even eligible for citizenship until 1942![1]

I taught Summer School, which was painful. Most of the kids attending hated school. Most had not done well in the regular year; I tried my best to make it fun, and by the end of the session all was well; but it was exhausting.

One year I taught outside of my district, at a school with a high Chinese population.

---

[1] https://www.history.com/topics/immigration/chinese-exclusion-act-1882

When I entered that classroom, it was full of junior high Chinese girls. Instantly, all talking ceased. One young lady raised her hand.

"Excuse me," she said, "We are here early because we want to learn. Can we please get started?"

I once asked a Chinese-American parent, where did such amazing respect for education come from, and she said words I will never forget:

"We may not always have food on the table — but we <u>will</u> have books."

In my experience, they are a surprisingly welcoming group, as perhaps others participating in a Tai Chi class can verify. The majority of participants in this ancient health exercise are Chinese.

But if you as a non-Asian want to participate, just do as I did. Walk to the back of any public class (beside a lake is a good place to find one) and imitate what they are doing. No one will say anything, but after the class they will probably approach you, and smile and ask if you want to sign up. When this happened to me, I thought uh-oh, here comes the bill — and there it was — one dollar a month. They used the money to buy batteries for the tape recorder (music) and an annual present for our teacher, international competitor Mei Chen.

Do Asian-Americans contribute, medically or scientifically?

17% of all American doctors are Asian, from only 5.4% of the population.[2]

"In the California biomedical field, Asian Americans and Asian-American Pacific Islanders make up nearly 1 in 6 workers (83,000). In the pharmaceutical and medical manufacturing industry, 1 in 7 (85,000); and 1 in 9 (100,000) in pharmacies."[3]

In the STEM professions (Science, Technology, Engineering and Math) Asians and Asian Pacific Islanders "make up 13% of the workforce, and many play critical roles in global research."[4]

---

[2] https://www.aamc.org/data-reports/workforce/interactive-data/figure-18-percentage-all-active-physicians-race/ethnicity-2018
[3] https://research.newamericaneconomy.org/report/aapi-americans-on-the-frontlines/
[4] https://www.pewtrusts.org/en/research-and-analysis/articles/2021/05/24/how-asian-american-and-pacific-island-researchers-are-contributing-to-the-future-of-science

When cure comes, we should remember their efforts.

As for the bullying and harassment currently being endured by the Asian-American community, such actions must be prosecuted to the fullest extent of the law, and labeled as the hate crimes they are.

And to the Asian-American community, I can only say, "You have enriched our country — you make the world a better place."

# 43 Of Crocodiles, and The Cost of Losing

I was on a freeway behind a big transport van, stacked double decker with cars. As I watched, pinned in by traffic on both sides, the van went over a bump. One of the cars wobbled in its chains, and rolled backwards — six inches. Then it stopped. But for a fraction of a second I thought it would roll off the van, and crush me.

Thursday, June 24th, 2020, was the deadline: to turn in signatures to get our initiative on the ballot. If by that date, 623,212 valid signatures had not been accepted by the California counties, and turned in to the Secretary of State, our stem cell renewal initiative would be denied, not even put before the voters.

If we were short by *just one signature*, Proposition 14 would die.

What caused that about-to-be-crushed feeling?

Something had gone wrong in Fresno. It had yanked all Prop 14 signature results.

How important was this? Did it lessen our chances to get on the ballot? Might it disqualify our $5.5 billion renewal for the California stem cell program?

I called Bob, but he wasn't home. I left a message, saying what I just told you: Fresno's signatures had been pulled, did this hurt our chances to get on the ballot?

$5.5 billion dollars…medically speaking, that might change the world. If the initiative (a) got on the ballot, and (b) won the approval of the voters, the scientists of California could take on the most tremendous challenges…

I will appear to digress for a moment, but am really not; you'll see.

What if you had to start your day — by grabbing a crocodile?

Not alligators, wrestling them is no big deal; I have done it. When I worked at Marine World, an aquarium-zoo in Redwood City, California, we had to move a dozen gators from one tank to another. The biggest one was about ten feet long, steam like dragon's breath coming from its mouth in the chill morning air. It made a noise like "HAAAAAHHH!

All I had to do was step across the gator's back, grab its jaws and hold them together. I lunged, closed the tooth-dripping jaws against the concrete tank floor, worked my fingers underneath, and held on. — easier than you might imagine. A gator's strength is in closing its mouth, not opening it.

Somebody wrapped tape around the jaws; that was it. We lifted the sluggish reptile like luggage, hoisted it onto a truck.

But a crocodile? That is different. Crocs are explosive, and tend to eat you.

A real-life monster, Gustave, the 20 foot long, one-ton beast of Burundi, is said to have killed 300 people. Some claim the creature has been killed, but I have not seen proof. Several years ago, he was observed to kill a full-grown water buffalo, dragging it from the bank and drowning it.[1]

So what is the connection between stem cells and a crocodile?

A scientist fighting disease is like a trapper hunting Gustave. There may be consequences.

If you fight the giant crocodile, and lose, you may die — as will perhaps more people, *future victims* of the reptile; but if you fight chronic disease, and lose — millions may continue to suffer and die.

What if you were a scientist wanting to cure cancer? How many lives might be saved, or lost, depending on the outcome of your struggle?

California held the future in its hands. We waited, for the counties which had not turned in their signatures.

The phone rang. It was Bob.

"Somebody made a mistake", he said, "but we are okay. We will probably get Fresno's votes, but even if we don't, we have a strong enough margin (San Diego's votes are not counted yet) that we should go over the top."

A few hours later the votes came in: both Fresno's, and San Diego's.

---

[1] https://en.wikipedia.org/wiki/Gustave_(crocodile)#Current_status

Here is the historic notification, (temphasis is theirs), followed by Bob's comment:

"Robert N. Klein, the proponent (of the initiative titled) AUTHORIZES BONDS TO CONTINUE FUNDING STEM CELL AND OTHER MEDICAL RESEARCH. INITIATIVE STATUTE, has filed more than 623,212 valid signatures with the counties.

"Therefore, pursuant to Elections Code section 9033, the initiative measure is eligible for the November 3, 2020, General Election ballot and all further signature verification can be terminated.

"On June 25, 2020, the Secretary of State will certify the initiative as qualified for the November 3, 2020, General Election ballot..."

(signed)

ALEX PADILLA, SECRETARY OF STATE

And the note from Bob Klein, marking the half-way point:

"Our coalition of over 60 dedicated patient advocate organizations fought to ensure our stem cell research Initiative. (We have) qualified for the November ballot, and we're eternally grateful for their commitment and vision," said Bob Klein, Chairman of Californians for Stem Cell Research, Treatments and Cures.

"During the past decade, California has made incredibly thoughtful investments and significant progress along our journey to developing therapies and cures, for diseases and conditions like diabetes, age-related blindness, cancer, Epilepsy, Parkinson's, Alzheimer's, and heart disease; it is critical to California families that this vital therapy development pipeline continue to be funded.

"Our state has always been a leader in medical and scientific research and therapy development, ranking second in the world when evaluated as a nation. Continuing to fund that mission is essential to the health of our families, stimulating economic recovery for California, with good paying jobs, created by this program."

"If Californians do not pass the 2020 stem cell Initiative, this vital research will come to a halt. Medical discoveries currently in the transitional pipeline will not be able to progress to clinical trials, delaying potential lifesaving and life-changing treatments for years, or even decades."

— Bob Klein, Chair, Californians for Stem Cell Research, Treatments and Cures

We had an official name and a number: Proposition 14: the California Stem Cell Research, Treatments and Cures Initiative of 2020.

We were on the ballot.

Now all we had to do was win.

---

[2]*https://ballotpedia.org/California_Proposition_14,_Stem_Cell_Research_Institute_Bond_Initiative_(2020)#Path_to_the_ballot

# 44 A Romance of Kidney Failure?

Everett Meyer is working to regenerate kidney tissue to save lives. (https://profiles. stanford.edu/everett-meyer)

An elderly friend of mine had end-stage kidney failure. His kidneys no longer filtered out the poisons in the urine for removal from the body.

Twice a week, his blood had to be taken out of his body and mechanically washed. Without this hemo-dialysis, the toxins would accumulate, and he would die.

His wife came along with him, and would wait in the car or the hospital hallway, reading a book or doing some knitting, while he had the procedure done. She did not like seeing her husband subjected to it; those needles poked into his arms, the blood pumped into a machine which washed and then restored it. It took a couple hours to clean perhaps 2/3 of his blood.

One day, on the drive over, he deviated from the usual route.

"Where are we going?", she said, "This is not the way to — "

"Shh, shh, it's a surprise," he told her.

When they pulled up in front of a residence, she was almost speechless.

"But — you know who lives here, it's — "

"Of course I know. It is your old boyfriend's house, the one you almost married many years ago — there were religious differences, your parents did not approve, I know all about it. I called him, told him we were coming."

"But why?"

"Because I love you and I want you happy. So, get out of the car, go visit with him, I'll pick you up in a couple of hours."

He watched her safely to the door, then drove off to get his blood washed.

Dr. Everett Meyer is a surgeon at Stanford, with a couple of decades of experience. He is the kind of doctor who, when cures do come, will be putting them into your body — also, he is the one you would want.

Dr. Meyer recently received a CIRM grant (CLIN2-11400) to work on kidney disease and kidney failure.

There is (right now) only one fully effective way to repair a malfunctioning kidney; and that is to replace the bad one with a good one. The problem is there are not enough spare kidneys available.

Dr. Meyer wants to do something different. Instead of replacing the entire donor kidney, he wants to just inject cells of healthy kidney tissue.

A major problem is rejection. The body does not "want" foreign organs or cells inserted, even for its own good.

To deal with the problem of rejection, immune-suppression drugs may be administered, lots of them, often for the lifetime of the patient. But these drugs brings their own series of problems; if the immune system is shut down, the body may not fight off infection.

Is there a way out of this dilemma?

What if kidney cells were used in their *progenitor stage* (not fully ready to function)? Would the body would be more likely to accept them?

That is what Dr. Meyer wants to find out. If he wins, it may revolutionize the field of transplantation medicine.

Organs like the heart, liver, lungs, kidneys and more might be repairable — with a simple injection of cells.

My friend is gone now, but I will always remember his act of self-sacrifice — and look forward to the day when such kindness is not needed any more.

# 45 Champions All

Behind every great struggle, there are people working quietly: sometimes little-known, but vital; here are a few of many, each of whom deserves their own book:

But first, key individuals in their campaign roles, which are sometimes different from their titles in the Americans for Cures Foundation:

Bob Klein — Executive Chairman

Robert Klein — President, Director of Strategy, Public Policy, Communications and Logistics

Melissa King — Deputy Campaign Manager, Head of Field Operations

Anna Maybach — Assistant Deputy Campaign Manager, Strategy, Engagement, and Communications

Jacqueline Hantgan — Assistant Deputy Campaign Manager, Coalition Building & Community Engagement

Mitra Hooshmand — Deputy Campaign Manager, Chief Scientific Officer

Don Reed — Public Media and Civic Group Outreach and Board Member

**JACQUELINE HANTGAN:**

My late father, Henry, my hero, sustained a Spinal Cord Injury (vertebrae C 6–7) when he fell in 2001 at a golf course. He has been in a wheel chair for the last ten plus years. It's difficult.

I became a stem cell research advocate after a conversation with Michael Goldberg in Utah. He was a congregant of our synagogue in Northern California. He talked about stem cells and Bob Klein and the initiative, Proposition 71. I had never heard anything about it, so I said, I like it, sign me up, who do I talk to?

Behind the scenes, people like Anna Maybach...

Jacqueline Hantgan...

and Melissa King do the hard chores to make the dreams of cure come true. (all three photos from www.americans.for.cures.com)

I worked with the "Three Amys", Amy Lewis, Amy Daly, and Amy DuRoss, Prop 71's gate keepers. I would do whatever work was needed. A lot of petition signing. Someone called my husband Yosh and said, "I don't know it is safe for your wife to collect signatures in the parking lot. She might get run over."

My political life began doing phone banking in New Orleans, where I attended Tulane University. My field was cultural anthropology and education. If there was a protest to be had, standing up for the right side of humanity, I would be there. We worked to defeat David Duke, Grand Wizard of the KKK.

My memories of prop 71? Lots of signature gathering: urgency, fantastic leadership, fighting to end suffering, Bob Klein who was perpetually busy. He had incredible capacity, running a full-time business as well as this, no extra hours. He was/is a man of incredible vision, altruism, compassion, selflessness.

I would arrange for speakers. Finally, Amy Daly said we have no more speakers, you do it. So I made my file cards and rehearsed with my husband and just did it. I was a utility player; I would call and find sponsors, endorsers, donors, faith-based advocacy, whatever was required. Tenacity is the key, be a polite pain in the neck — we do not have to take no for answer.

(Don — "Jacqueline uses the FU3 policy — follow up, follow up, follow up!")

We had our own heroes growing up, now it's our turn to accept the challenges.

**MELISSA KING:**

Melissa is a terrific athlete, an avid open water swimmer, and a long-time member of the Dolphin Swimming & Boating Club in San Francisco. In 2011, she swam on the first all-women relay from the Golden Gate to the Farallon Islands, the Maui Channel relay as part of an all-women fundraising team (to help the California Marine Mammal Rescue Center) and also a birthday celebration relay with fellow Dolphin Club members across the Catalina Channel. She has completed the Alcatraz swim many times, swum the length of the Golden Gate bridge more than once and has swum in open waters as far flung as Hawaii, Mexico, Ireland, Spain, Finland and India. A lover of winter swimming, Melissa has twice been the first to complete the Dolphin Club's annual Polar Bear Challenge, completing 40 miles of swimming in just the first several days of the three months allotted for completion.

Of her stem cell efforts, Melissa says: "I headed up Field Operations, including leading the effort to get volunteer-driven signature gathering accomplished under COVID conditions. My work was central in mobilizing our coalition of patient advocates, scientists and others — developed in the years leading up to 2020 — throughout the campaign: for signature gathering, outreach, communications and the final big push in November. I created the Patient & Medical Advocacy Committee (PMAC), including structure, membership, workflow and output of this crucial group of dedicated volunteers through the election."

In addition to working on Prop 14 in 2020, Melissa reports serving as a " member of the successful 2004 campaign for California's Proposition 71, which created the California Institute for Regenerative Medicine (CIRM) and secured its first $3 billion in funds. During the final weeks of the Prop 71 campaign, at the request of campaign Chair Robert Klein, Melissa worked with legal counsel to form the non-profit that is now Americans for Cures Foundation.

Melissa played a central role in the founding, building and first seven years of operations of CIRM, the unique $3 billion State research funding agency created by Prop 71. As Executive Director to the Board, she worked hand-in-hand with CIRM's Chair and its 29-member governing Board on setting the strategic vision for the agency as well as developing the processes and policies by which CIRM still operates. Ms. King guided the Board through 1500+ hours of public meetings, through decisions including where CIRM's headquarters would be and who would get the very first several rounds of grant funding, as well as developing and finalizing the CIRM IP policy and the Medical and Ethical standards by which CIRM researchers must operate.

## ANNA MAYBACH:

"My first scientific involvement came when I was a child in Colorado. We spent much of our Summers in our cabin with no TV; but we did have a rainy-day science projects book.

At Vanderbilt Children's Hospital, I worked with children with terminal illness. The social neglect they faced when it came to discussions about their own palliative care process precipitated my research into the subject, leading to the development of a vocational program in the hospital that taught providers how to include children in their own palliative care planning.

I have been ocean diving since I was 12. My first dive was in Tahiti, where I saw sting rays and reef sharks. My best shark diving? At end of the dive I was just floating and the dive master threw me a snorkel-mask and said "Look down!" There was a whole bunch of sharks (tigers, lemons, reef sharks) circling below.

At Columbia University, I met Alyssa, Bob Klein's daughter, and Bob himself.

With Prop 14 imminent, I applied at Americans for Cures Foundation. My role focused on research and strategy, content creation of major documents, data collection and analysis, developing regional fact sheets, and disease fact sheets.

My favorite "Bob memory"? It amazes me how he can fascinate people from every walk of life, a great speaker, captivating, for non-scientists or scientists alike.

## ROBERT "ROB" KLEIN:

My first memory of stem cell research was probably at a JDRF conference. Dad was talking about Type II diabetes, which my brother Jordan had. I did not understand the science at first, but just knew it was important.

During the campaign, (Prop 71) I remember the campaign office, jammed with people working: frantic, cheerful, dedicated. In 2007, I first sat in on the CIRM board meetings. All the science background I had was one freshman year science course, Biology 101, so I learned hands-on. As I became familiar, I became fascinated; you cannot learn about stem cells without becoming excited.

I have two jobs: one is with the family business, Klein Financial, where I work on asset management and the future development of multiple multifamily apartment housing projects, which always contain an element of affordable housing, so folks of modest means can have a decent place to live. At Americans for Cures Foundation, I work on political analysis, as director of Government Affairs.

(Note: during the Proposition 14 effort, Robert served as Campaign President. RNK, personal communication.)

Need for Proposition 14: it became clear the opposition would continue trying to block some of the most powerful forms of research. Using court cases and legislation, they were continuing an effort to push back research. We needed government to either support the research, or at least get out of the way.

It was overwhelmingly clear that a second initiative, Proposition 14, was necessary, especially considering all of the amazing progress that had been made, *much of which could be lost* if additional funding was not approved.

Even research showing promising results in mid-stage human trials was threatened, as private funding sources were just now figuring out how to monetize these kinds of therapies, how to produce value for investors who might otherwise look elsewhere.

Additionally, the traditional pharmaceutical approaches to treating disease are much more familiar and predictable; it will take a number of breakout therapies and/or cures to motivate private industry to dive deeper into what appears to be a much more promising field of research long term.

There were (and are) special problems to be faced. Eventually, the model of medical payments has to change. The standard procedure now is a lifetime of continual payments, compared to cure, which may be a one-time payment…Patients should not have to wait for private capital to decide how to adjust their payment models. Morality demands that if we can advance toward lessening suffering, we should do so. Once these cures are developed, demand from the public and their elected representatives will be intense enough for working private funding models to be developed and instituted quickly. For that to happen, the promising CIRM-funded research and clinical trials underway need support to survive past the "Valley of Death", when funding is so limited, and also receive FDA approval.

We have seen continued challenges to embryonic and fetal stem cell research, like the effectual ban on funding for fetal stem cell research the Trump Administration instituted, regardless of the highly promising results that research was having. Additionally, highly ideological members of Congress continue to push to reduce research funding, or ban it outright; putting forth legislation to do so every cycle.

When drafting Proposition 14 and campaigning for its passage, we (and everybody else) had no idea who would end up occupying the White

House and leading Congress. Fortunately, a majority of those occupying these positions now (in 2021) do support this critical research. But we have no guarantees that this will remain the case in 2022,or in 2024 and beyond. The efficacy of medical research and the speed of its progress can be immensely reduced when tied to the partisan tug of war that today defines our democracy.

Patients and their families should not have their suffering extended for years due to political strategies and/or the shift of ideological politics every 2–4 years, on both federal and state levels. This research requires consistent, uninterrupted funding, and the California Institute for Regenerative Medicine was and is the best opportunity with the most capacity to provide that funding.

Early polls showed us with a strong lead, much like the 20 points we won by with Prop 71. Then came COVID-19, which really threw us a curve. With public signature-gathering severely hampered, the game had changed. With State economic problems, forest fires and more, people were understandably worried about cost, and the opposition took every opportunity to mislead them into thinking that the budgetary impacts would be devastating: trying to make voters think they had to choose between Prop 14 and other immediate spending needs like housing, homelessness and environmental concerns.

Fortunately, our coalition of patient advocates was strong. They backed us not just with endorsements, (which were important), but with work: individuals printing petitions at home and then safely reaching out to their families, friends and co-workers, and door to door, fighting for every signature. We eventually got all the valid signatures we needed, with a healthy margin.

But when we were actually on the ballot, gathering funds for the campaign was complicated by major "donor fatigue". We knew we had a compelling message, but how could we get it to the voters? Regardless where donors were on the political spectrum, they were all but tapped out by the Presidential, Senate, and House elections. This impacted us greatly. A donor who might previously have supported our stem cell effort with a million dollars might only chip in fractions of what they gave before, if they were able to contribute anything at all.

We refused all funding from Pharma, Biotech and economically interested entities who might have stood to gain financially from the

Initiative's passage. It was immensely important that voters know that this was not a giveaway to those with financial interests, and that our core motivation was to lessen the suffering of patients and families.

Bob himself had always been the major donor to the first campaign; this did not change. Toward the end of the campaign, Bob and his wife Danielle worried that our projected margin of success was too slim. They did everything they could to provide additional financial support, to push us across the line.

If we won, there were terrific possibilities. California scientists working with CIRM grants were fighting cancer, paralysis, Huntington's disease, diabetes, Parkinson's and more — conditions and diseases long considered incurable. In addition to projects already begun, we hoped to extend our reach, giving informational support to other states.

While every research dollar must be spent in California, we needed to work internationally, involving the best scientists, not only in our state and nation, but in the world.

CIRM is fighting afflictions that plague all mankind, regardless of state and national boundaries. CIRM hopes to enter into collaborations with the best minds from all over the world — to significantly accelerate the race to develop more effectual treatments and cures for all.

# 46 Mud and Dust, Fire and Flood

Nature is worth preserving: author and the greatest swimmer in the sea. (*Notes from an Underwater Zoo*, Don C. Reed, 1981)

If you ask Bob Klein what is most disturbing about science reporting, he will answer that there almost isn't any. Too many newspapers are gone, and with them, their science reporting. The great news channel CNN?

They "re-organized" and eliminated much of or their entire science departments.[1]

There is also the change in the readership, meaning us. How many people are willing to take the time to read a lengthy science article?

Frighteningly few.

This matters, and not just for stem cells.

You cannot have escaped noticing Earth's increasing ill health. Streams and rivers we played in as children, now are dry. Summers, once gently warm, are now blasting heat, so we choke without air conditioning; Winters freeze like dry ice; animal species — so many! — are just gone. And if the littlest creatures die off, what happens to their predators, and those who eat them?

Mostly, what happens is gradual, unnoticed, not bolts of lightning with background music.

But every so often we get a hint of what may happen — if Earth strikes back.

DUST: One blasting Summer I was in Texas, and it was just miserably hot.

I was crossing a dry brown field, when in the distance I saw something coming, towering and strange. It was like the movie, THE WIZARD OF OZ — when the tornado — "It's a twister!" — picks up Dorothy in her house.

What approached me now was what Texans would likely dismiss as a "dust devil" — a miniature tornado. It might have been a hundred feet tall and twenty feet across, sucking up stuff as it came.

If I stayed where I was, it would pass me by. But I would never get this chance again.

I ran full-speed toward it.

There was just time to think, "this might not be such a good idea!", and then I was inside the shrieking vortex.

It was hotter inside the swirling wall of raised dirt. A million fingers clutched at me. I squinted my eyes, saw dust and filth and twigs. There was an instant of stillness, inside the eye — then the roar again. The wind subtracted all details of sound — all I could hear was a roar. And then the dust devil moved past, leaving me drenched with dirt and sweat and mud, and cockleburs all over my clothes.

---

[1] https://www.aps.org/publications/apsnews/200904/journalism.cfm

FIRE: You know the fire season of California, so dry the smallest spark ignites a blaze? Have you driven down the freeway, when walls of orange flame come roaring toward the road. It's far away, but is it far enough? And if it comes too close, will the fire leap across, or ignite the cars? Anything seems possible, and oh, the traffic moves so slowly, as you pray it will not stop.

I remember once trying to light an old-fashioned gas stove. My cousins and I were all city boys, and none of us knew how the thing was done. I figured it just needed a little more gas. So I turned the knob all the way over, waited a moment and then scratched the wooden match — PHOOOOFF!!!

My eyebrows were gone. Also, I was partially bald. What foliage I retained on my head was just frizz.

But I did not lose my sight; the building did not burn down. So I counted that as a learning experience.

FLOOD: not far from where I live is a channel, which carries rainwater from the hills down to the sea. It is dry, now, mostly.

But when we first moved in, there was a wonderful three-day rain, the kind we used to have routinely, long ago. The rains came thumping on the roof, pounding like devils, trying to break through.

I went outside, taking a walk in the rain, making up little poems about "roaring, pouring torrents of the storm", loving how we were all connected by the rain, engulfed with it. I walked down to the channel, wondering how high the water level was. It was right at the edge. Another few inches and it would come over the top, into the neighborhood. We would need sandbags, but there were none.

Later that day I had to drive under a road bridge. The water was hubcap-high; there was some doubt if my car would get through. Would it flood the engine?

And then the rain — stopped. That's all, it just quit falling, leaving me with a question I did not want answered: what would a real flood be like?

MUD: I had a hint of an answer one day, the night before Marine World opened its newest attraction: WHALE-OF-A-TIME-WORLD, a children's playground.

Part of it was a Tarzan swing over the canal, winding peacefully around the park. To keep people from getting wet when they fell, (which they would) there was a water-tight rubber mat underneath, connected to large padded boxes, blocks of styrofoam.

The boxes rested on mudbanks. Someone had to burrow into that mud, and fasten some bolts, on the underside of the boxes, so they would stay together. I was head diver then, so it fell to me.

How hard could it be, I remember thinking. Just fasten the straps of my hookah hose, attach the other end to an air tank, go underwater in the canal, dig a little tunnel in the mudbank, then crawl underneath on my back.

It did seem simple, at first. I had a chisel, a mesh bag of washers and nuts, a wrench. I started off with a shovel, which I soon discarded, just used my hands to dig a hole in the wall of mud. Then, lying on my back, I fumbled underneath the Styrofoam, fingers seeking till I found a bump. The bump meant a bolt. I chiseled the foam away till I reached the bolt, slipped on a washer and a nut, snugged them up tight. It was intensely claustrophobic.

At midnight, I was still not done, and the attraction was due to be open tomorrow. The manager of the park, Mike Demetrios, was standing by, gently pressuring me to get the job done.

Exhaustion set in. Water stole my body heat. The tunnels I made in the mud now were just barely wide enough to wiggle in.

I was all the way under, working on the last bolt, when it happened.

The mudbank collapsed around me. I did not realize what had happened, at first; but I could not move. I was entombed in absolute dark.

I shifted my head; mud oozed under the lip of my mask. I realized no one topside would know I was in trouble. I would just quietly run out of air. When the tank was empty, I would die in the mud.

Then I thought: they will dig me out, eventually, and when they do, one of the shovel edges might bite into my shinbone. For some reason that thought irritated me so much, it gave me new strength. I used the sides of my legs and hips to pull, side to side, did I get a fraction of an inch? Fighting for traction, nothing happening, no progress — then something a little bit, perhaps — then more and more and an audible SHLOOOP — I slipped out into the dark water.

Of course, I still had to go back and finish the job, but I made the tunnel extra wide.

When I finished, Mike Demetrios handed me a bottle of Southern Comfort, saying, "I don't know if you are a drinking man," he said.

"If I wasn't, I would be now," I said, giving it back to him empty.

I haven't had a drink in thirty years, (except something vile called a "jello shot" which my daughter-in-law Terri conned me into eating), but I remember that "Southern Comfort" with appreciation. It made the long walk back to the changing room positively — bearable.

What can Earth do, if she really "gets mad"?

Look around.

See the California drought attacking our state, or mudslides roaring through a city in Japan, or Burmese pythons taking over the Everglades — or when the ocean picks up a giant boulder, and throws it at a lighthouse, smashing off its top — and I don't even know what happened to the lighthouse keeper!

All I know for sure is that when Earth strikes back, it will seldom be the way we imagined it — and it will never be convenient.

# 47 Nearing the Top of the Mountain

Nobel laureate Shinya Yamanaka sums up why California's stem cell research program must go on. (www.ucsf.edu)

There was a couple in Vietnam who lived in a cave, their home and village having been bombed out during the war. The husband was a quadriplegic, paralyzed in upper and lower body. The woman was a quiet heroine.

Every morning she would grab her husband under his arms and drag him outside, lean him up against a tree, so he could watch what was going on during the day.

Then she would go out and try to hunt them up them something to eat.

In every nation there are families devastated by disability or disease; most of their stories we will never know.

But we are not excused from our human obligation, which is to help. The question is — how? I cannot personally bring the cure for paralysis to that family.

But I can work so there will one day BE a cure.

For those who have a paralyzed family member (like my son Roman) please know that the California stem cell program has spent more than $52 million fighting spinal cord injury. To see exactly where the money went, go to:[1]

Newly paralyzed young people have been helped by CIRM-sponsored stem cell therapies. Further research is needed to help those with long-term (chronic) injuries.

That is just the tip of the iceberg.

Let me tell you a story, which is not my own. It is a Walt Disney cartoon: "THE RESCUERS".*

It begins in the United Nations, showing all the different countries, cheerfully working together, which of course I love.

But in the basement of the U.N. is another civilization, of mice, featuring the voices of Zsa-Zsa Gabor and Bob Newhart. They have a little song, "R-E-S-C-U-E, Rescue Aid Society" which echoes in my mind today.

But the whole point of this beautiful little story is that the mouse United Nations sends out two representatives — to save the life of an endangered child...

Make it a point to see the film. It is fun, of course, but there are moments in which I genuinely have to choke back tears — because saving lives is CIRM's reason for existence.

The California program has saved the lives of 50 children, all diagnosed with the "bubble baby" disease. This was previously a death sentence, "Severe Combined Immunodeficiency", or SCID. Those children, including our small champion, Evie Vaccaro, are alive today.

Neither did Brenden Whitaker die, though he had the (usually fatal) granulomatous disease. He is going to become a doctor, which now he has the years to do.

---

[1] https://www.cirm.ca.gov/our-progress/disease-information/spinal-cord-injury-fact-sheet

*I hasten to add that neither I nor CIRM have any connection to the Disney company. But I love their film with its gentle message, and thank them for it.

Remember the baby who had thalassemia, but was operated on by UCSF's Dr. Tippi Mackenzie inside his mother's womb — first in the world — and whose life was saved, because of the actions of real-life "rescuers": scientists and doctors.?[2]

Today, two FDA-approved treatments to fight blood cancer are available, thanks to Dr. Catriona Jamieson's CIRM-funded work at UC San Diego.

Over one hundred clinical trials are underway or completed: 70 trials California directly paid for, and an additional 30 in which the research was funded earlier.[3]

But after 16 years of carefully funding research, CIRM was running out of money. If we did nothing, the program would die. This must not be allowed.

But don't take my word for it — I am not even a scientist!

Listen, instead, to three of our world's most brilliant people — all recipients of the Nobel Prize, Earth's highest honor, for their contributions to science.[4]

**Paul Berg:** "The creation of CIRM was a bold initiative by the citizens of California…an exceedingly promising scientific breakthrough…

"(CIRM-funded) scientists achieved world leadership…exploring opportunities for human cures or treatments. Many clinical trials for treating a wide variety of human afflictions are in progress…now is not the time to slack off support… *We are near the top of the mountain*, but we need continued support to reach the pinnacle — providing cures for the ills that still plague humankind."

**David Baltimore:** "CIRM money has allowed California to be in the vanguard of stem cell research…In this new and highly promising area of research, CIRM support has been wide-ranging…a model for the country."

**Shinya Yamanaka:** "CIRM is a global leader of stem cell-based basic research and clinical application, providing great hope to patients with intractable diseases. I believe CIRM's initiatives will deliver innovative therapeutic options to solve the patients' most pressing needs.

---

[2] https://www.cirm.ca.gov/our-progress/awards/utero-hematopoietic-stem-cell-transplantation-treatment-fetuses-alpha
[3] https://www.cirm.ca.gov/clinical-trials
[4] CALIFORNIA CURES, by Don C. Reed, World Scientific Publishing, Inc.

"The California stem cell program (CIRM) has been helpful to many scientists during a time when obtaining funds for research has been increasingly difficult. Without such funding, research cannot go forward.

"It is my hope that the people of California will choose to continue this useful institution."

# 48 Supply to the Eyes?

Dennis Clegg battles the most common form of blindness — (Age-related Macular Degeneration) in a new way. (blog.cirm.ca.gov).

November 28, 1949: August Rauhut, a veteran of the Spanish-American war, climbed up on the north end of the Golden Gate Bridge, and leaped — becoming the 116ᵗʰ suicide from the Golden Gate's span.

He was the brother of my great-grandmother Christine Rauhut. He was depressed from what was then described as "failing vision", a fear that he was "going blind" from what was probably Age-Related Macular Degeneration…

— information courtesy of Will Snyder, family historian, author of "Lockedhart" — True Tales of Early Fall River Valley, Natives and Pioneers 1849–1868.

Somewhere in time, probably early Summer, 1952. I was 7 years old, standing in front of my house, in Berkeley, California.

There were almost no cars on the street. Families might have one, but no more than that. Mostly, we took the bus, or walked. With so few cars, there was much less pollution. Also, I think it had just rained.

But what I remember is the color of the sky. It was the most amazing shade of blue. No clouds for distraction, just that astonishing, glorious, deep deep blue... I raised my skinny arms and tried to absorb all that beauty...I was transfixed by it, mesmerized, I would stand there forever.

"Don-*ald*!" My mother yelled, leaning on the second syllable of my name, as she did when there was trouble.

"What?" I answered, not taking my eyes from that amazing sky. I stood on my tiptoes; I could almost lift off and fly.

"Dishes!" she said. I heard the screen door open and shut with a bang.

"I'm looking at the sky!" I explained hurriedly.

"It will be there tomorrow," she said...

She was both right and wrong. The sky was there in the morning, of course.

But I never saw that particular shade of blue again — and I have looked for it, head back till my neck ached, searching for that one exquisite hue of sky — what color was it? Aquamarine? I have no proper name for it. The closest I ever saw was a billboard for mint-flavored Lifesavers. It was a painting of an old wooden bucket in a quiet stream. The bucket was half-full, and the sky reflected on the water: it was almost that shade of blue, but not quite. It is in my brain; I wish I could share it with you.

But still I have my sight, and at 75 years that is no small blessing.

In California alone, an estimated 450,000 senior citizens (half a million!) are not so fortunate. They have Age-related Macular Degeneration (AMD), the most common form of blindness for the old.

AMD begins as a little dot in the middle of your vision, a spot you cannot blink away, nor rub out with a washcloth. The dot grows, big as a thumb in your eye — and after a while, the curtain falls.

What happened?

At the back of the eye is a layer of cells called the Retinal Pigment Epithelium (RPE). This is a "supply store" for the eye, providing nutrients for light-sensing cells. If those cells grow weak, so does your vision; if enough die, you are blind.

And that is why the California stem cell program exists, to fight such horrors.

One of my favorite CIRM awards is the Disease Team grant. As the name implies, it is a team effort, a dream team, a gathering of champions.

If you follow research on eye disease cure, there are names you will recognize, like: Amir H. Kashani, Mark Humayun and David Hinton of the University of Southern California, Dennis Clegg of the University of Santa Barbara, and Pete Coffey of the United Kingdom: each a leader in the field.[1]

Put them all together, and that is a CIRM disease team.

What is their method of attacking AMD?

Replenish the eye supply store!

"Effective treatment could be achieved by proper *replacement of damaged RPE...cells* with healthy ones."

To continue the store analogy, you need shelves to stack the supplies, right?.

So, at the back of the eye, the scientists want to put a tiny mesh structure, (the "shelves") on which go the replacement cells. This "mimics the natural membrane, ...important for the survival of the RPE."[2]

And then — you stock the shelves with stem cells!

Here is Dr. Dennis Clegg:

"Our part of this project was to differentiate hESC (human embryonic stem cells, otherwise discarded) into retinal pigmented epithelial cells, seeded onto a synthetic membrane and implanted in the back of the eye — to replace defective RPE cells.

"We were able to rapidly build our stem cell program via recruitment of Peter Coffey from London...and Jamie Thomson to a partial appointment at UCSB. Pete, Jamie and I serve as Codirectors of the Center for Stem Cell Biology and Engineering, which continues to thrive.

"I am also co-founder of a startup company called Regenerative Patch Technologies, which is carrying out the clinical trials under the leadership of Jane Lebkowski, who was VP at Geron during the first clinical application of hESC." — Dennis Clegg, personal communication.

Did you notice the names of Jamie Thompson and Jane Lebkowski? These are home run hitters! Jamie Thompson is the American scientist

---

[1] https://www.cirm.ca.gov/our-progress/awards/stem-cell-based-treatment-strategy-age-related-macular-degeneration-amd
[2] Stem cell based treatment strategy for Age-Related Macular Degeneration (AMD), Grant # DR1-01444

credited with the discovery of human Embryonic Stem Cells (hESCs), and Jane Lebkowski oversaw the partially CIRM-funded Geron clincal trials, which sprang from techniques developed by Hans Keirstead's work on a Roman Reed Spinal Cord Injury Research Act grant, way back in 2002!

Several foundations helped with the funding for this work toward curing AMD, too many to list here, but the big gun was CIRM, which provided nearly nineteen million ($18,904,916) dollars.

What happened next? Before I tell you that, remember how clinical trials work. The first one is always for safety alone. It must be proven that the therapy will do no harm. Only then will the trials go on with larger amounts of the therapeutic agent, to measure its efficacy, meaning does it work.

Of the five people who started off in the trials, none were made worse.

Also, the disease *stopped going the wrong way.* Remember AMD is a progressive disease, meaning your vision gets on a downward path, worse and worse until you go blind. If you are stable, that means you are keeping vision you very likely would have lost. Also, one person's eye exam proved substantively better, improving by 17 letters on the eye test.

So what happens next?

"We are very excited that some patients in the phase one trial have experienced improvements in vision.[3] and are gearing up for a phase 2 trial."

They are moving ahead;[4] the vision of millions is at stake.

It is to be hoped that our grandchildren will one day see that perfect sky — and I will get to long continue — watching my grandchildren!

---

[3] https://stm.sciencemag.org/content/10/435/eaao4097.abstract
[4] https://www.cirm.ca.gov/our-progress/awards/stem-cell-based-treatment-strategy-age-related-macular-degeneration-amd

# 49 How Roman Gets Up in the Morning

<u>Sam Maddox's</u> book, *Quest for Cure*, offered what was for me the first glimmer of hope that paralysis might be alleviated. (https://scifirst90days.com/about-us/meet-sam-maddox/)

If you meet Roman once, you will not forget him: this burly golden lion of a man. Although a quadriplegic, paralyzed from his shoulders down, he is very independent, living largely on his own.

Many times, he has driven his adapted van across the state (and the country!) to counsel a newly-paralyzed person, or help a group to start their own spinal cord injury research program.

But everything is much harder now. He can still put himself to bed, but it is increasingly difficult, and I am glad when his son Jason helps with the transfer.

Champion   scientists   HansKeirstead...   (https://aivitabiomedical.com/about-us/our-leadership/hans-s-keirstead-ph-d/)

...and Mark Tuszynski (www.eurekalert.org) advanced the possibility of paralysis cure: first with small grants from the Roman Reed Spinal Cord Injury Research Act, then with more substantial grants from CIRM.

Because of the number of years (twenty-six) he has spent in a chair, Roman's skin is breaking down. He has pressure sores, shallow wounds on heels and hip, and they never completely go away.

Every morning, I go to his house and treat the wounds as best I can: cleaning, putting on medical honey or various salves and sprays, changing the bandages.

His dog Dartanya (a 200 pound French mastiff) likes to stay inside guarding Roman, but she also loves her walks. To get her out of the way, she and I have a deal. My part is to take her for a walk, down and around our long city block, (she tugs violently at her leash, dragging me the entire journey) after which she must stay in the back yard (without complaining) while I tend to Roman.

When we are through, and Roman is back on his chair, Dartanya returns.

We made one all-out effort to end Roman's pressure sores, keeping him off the chair (in bed, which he hated) for several weeks. The doctor said he had sepsis (body poisons) and a tube was inserted through his arm, into his heart. His former wife Terri was kind enough to come to the house every night and change the tube.

If the condition got better, the thinking went, he might qualify for a "flap" surgery.

Three times a day I went over to his house, to get him food and clean him up.

The wounds got better, improving so much, they were almost gone.

The tube into his heart was removed, Roman went back to his chair — and the wounds immediately returned. Apparently the skin was now too thin to take the pressure.

So how does a paralyzed person get out of bed?

First, he gets dressed: bandages changed, fresh clothes on, while still on his back or side.

When he is ready, I shift his legs off the bed with my left arm, while his right arm interlocks with mine. In one swift motion, he is sitting up. I slip his shoes on.

His chair is positioned nearby. I move it closer to the bed.

I grab the trouser waistband from both sides, wait for his command, "1-2-3!"; we combine our efforts: two moves, shifting him first to the bedside, then to the chair.

I am 75 years old. Roman weighs 220 pounds. Fortunately, I am an old-school weightlifter, (not bodybuilder) and know how to use what strength I still possess.

But do you see why Proposition 14 is so important?

There are millions of families like ours, struggling to cope with paralysis, or other forms of chronic disease or disability.

If the voters say YES to California's new initiative, Roman has a chance at cure, him and millions like him.

There are no guarantees, of course. But if the funding is there, we have a chance. Without the money, no chance: we are stopped.

Dr. Hans Keirstead said of CIRM, "This kind of opportunity comes maybe once a century."

There is no choice, no fallback position. We must win.

And there are two men I would bet on.

First of course is Keirstead himself. He does amazing work; remember the five individuals who recovered a measure of upper body function after an embryonic stem cell therapy? Their recovery was based on Keirstead's work, originally funded by "Roman's Law".[1]

But Keirstead has many areas of challenge, including trying to cure cancer and fighting suffering in impoverished countries. He brings vitamins and medical equipment (including hundreds of donated wheelchairs) where it is most needed, including African countries. He was recently honored by the Canadian government for his work fighting COVID-19.[2] His attention is divided, but genius is unpredictable.

Mark Tuszynski is the quiet opposite. His energy is focused on one field of study: spinal cord injury and other forms of neurological disfunction (Alzheimer's disease especially) and he has been fighting to cure them for almost 30 years. He is calm, determined, focused: unstoppable.[3]

If either or both of these champions wins, so does the world.

I have a recurring dream, in which I go to Roman's room to get him up. But the bed is empty, he is not there — did he die?

And then I hear a voice, "Hi, Dad!" and feel a hand on my shoulder. I turn, and it's Roman, on his feet, standing, un-assisted, his full six feet four inches towering in the room. He smiles, and raises his hands, like "Check this out!"

There is a great verse in the Bible, "Without a vision, the people perish."

— Proverbs 29:18

That is my vision, my dream; that Roman is on his feet again — and I don't have to worry any more.

---

[1] https://blog.cirm.ca.gov/2017/01/25/good-news-from-asterias-cirm-funded-spinal-cord-injury-trial/

[2] https://www.biospace.com/article/releases/aivita-biomedical-ceo-hans-keirstead-named-to-20-in-2020-by-consulate-general-of-canada-for-contributions-to-covid-19-research/

[3] https://neurosciences.ucsd.edu/centers-programs/neural-repair/index.html

# 50 A Scientist's Life for You?

Bucky Fuller's geodesic dome is the world's most efficient housing structure.

If you want to be a scientist, do not skip pages two to four of this chapter. The first page and half is fun — but the last part could make a difference in your life.

Now. In my garage is an odd-looking sphere: pencil-thin sticks and wires all connected. If you push on any portion of it, the entire structure

Want to be a stem cell scientist? Know someone who does? Dr. Kelly Shepard has a program you should know about. (www.stemcellpodcast.com)

squeezes together. It is delicate but strong, and very light, as though it might lift up and fly away.

It took me most of a Summer to put it together, a 3-Dimentional puzzle: a geodesic dome. Invented by R. Buckminster Fuller, it is said to be the most efficient way to enclose space: a house designed on geodesic principles can reportedly be built with one per cent of the materials needed for the usual house.

Think of France's Eiffel Tower, built long ago by Gustave Eiffel. The similarities are plain. The geodesic dome and the Eiffel Tower are made on *suspension principles.* (Stay with me, we get to the part about you in about one minute!)

If you take a piece of string and tie it to a weightlifter's iron plate, you can lift it up — the weight dangling in suspension. But reverse their positions? if you take the weight plate on the same piece of string, and hold it over your hand, and let go — what will happen? The plate will fall, maybe break your hand. That is *compression*, a not — efficient way of supporting weight. You would need a thick string indeed to hold up the weight from underneath!

Suspension, compression…

Compare two famous buildings, one built on suspension principles, the other on compression.

The Eiffel Tower is very light and tall; at 1,063 feet in height, it weighs ten thousand tons. (The air inside it weighs almost as much as the metal!)

The Washington Monument is half as high (555 feet) — and eight times heavier (eighty-one thousand tons).[1]

It has been said (and I have no idea if it is true) that you could lie underneath a corner of the Eiffel Tower, and there would only be about 150 pounds pressing on your chest, maybe the weight of your brother. But if you tried the same experiment with the Washington Monument, there would be several tons squashing down — which would be uncomfortable.

I wrote a letter to Buckminster Fuller, expressing enthusiasm for his genius in architecture and world unity.

For which I received a form letter, saying, in essence: "I get hundreds of letters and cannot answer them all. If your ideas are of any value, they will eventually be published. When they are, write me again, and I will answer."

So I did — and he did. He kept his promise. I sent him a copy of my first book, NOTES FROM AN UNDERWATER ZOO, to which he responded with an autographed picture, which I still have, and a note, saying. "I like your book, and I like you.": a small act of kindness, which has echoed through my life.

I met him once, at a book signing. He was very short, and broad across the shoulders, like a miniature weightlifter. His wife was with him, and the affection between the two was clear. His grip was strong. His hair was white and bristly.

I reminded him of the incident, of which he plainly had no recollection — but he kindly volunteered to write it on my book. But there was a queue forming behind me. His wife nudged him, and I said no, just his signature would be great. He scribbled it with a felt point, and I still have it.

Not long after that, his wife of fifty years became ill. Bucky immediately canceled all his appearances, got a chair and sat beside her — until she died.

Shortly after that, he too laid down, and joined her in the great beyond.

Genius, of course, cannot be taught. **But a career in science is quite possible.** If you get a break or two, that is — so get out your

---

[1] https://www.uh.edu/engines/epi1608.htm

phone or pen and paper, and copy this URL www.cirm.ca.gov That is the headquarters of the California Institute of Regenerative Medicine: a website to explore.

Here is the question: would you like to become a stem cell scientist, or do you know someone who does?

If so, this article is a gift for you; or, more accurately, *three gifts* all from the California Institute for Regenerative Medicine (CIRM).

**First**, if you are in High School or college, your science teacher might want to teach a stem cell program, If so, there is a free curriculum ready and waiting for her/him to teach. This was developed with the cooperation of CIRM as part of Senate Bill 471, (Romero, Steinberg and Torlakson), the California Stem Cell and Biotechnology Education and Workforce Development Act of 2009.

It offers a wide variety of teaching ideas: "On this site, you will be able to: explore opportunities for teaching your students about stem cell science; find resources for student projects; locate a scientist to discuss stem cell science with your class; get involved in the growing areas of stem cell research and biotechnology."[2]

**Second**, for high school students, here is an amazing possibility. Would you like to spend a Summer working in a stem cell research lab? If so, check out SPARK: the Summer Program for Advancing Regenerative Medical Knowledge.[3]

SPARK "offers California high school students an opportunity to gain hands-on training in stem cell research at some of the leading research institutes in California. The program specifically selects students who represent the diversity of California, particularly those who might not otherwise have opportunities ... due to socioeconomic constraints."

**Third**, the BRIDGES program may help you answer the age-old employer question: "what experience do you have?" Nobody wants to say "How can I get experience if nobody will hire me?" Instead, how about: "I worked all year at a worldfamous biomedicine lab — and got paid $40,000 for it!"

Here's how it works: college students take basic stem cell training at their home college for the first year, followed by a second year in a lab

---

[2] https://norecopa.no/3rset-resources/model-stem-cell-curriculum
[3] https://www.cirm.ca.gov/our-impact/internship-programs

position at one of the research institutions taking part in the program. For this they will receive a stipend, up to $40,000, as well as priceless experience working at the lab.

Across the state, at least 14 colleges are taking part. I visited one of these, San Francisco State University, meeting Dr. Carmen Domingo, who directed the program there. She was bursting with pride and enthusiasm, and her students were the same. They were going through their first year of the stem cell program. They would spend the second year (paid) at one of the partner institutes.

According to Dr. Domingo, nearly 70% of the young students were either first in their family to achieve college, or a member of an under-represented minority. They could take a Master's degree for scientists on a Ph.D. track, or a professional degree for lab technician and other related careers. How did they feel about it?

"My best friend suffered a spinal cord injury and was told he would never walk again...my grandparents died of Alzheimer's...my father-in-law with early onset Parkinson's...stem cell research was the one field that might help them all. The CIRM program gave me (a) foot in the door..." Ian B.

"The California stem cell program literally changed my life. CIRM funding, through its focus on lower-income students, allowed me to pursue a Master's program, providing me with the tools necessary to broaden my knowledge and understanding of the science behind stem cells...playing an integral role in forging my career in stem cell research advocacy..." Yimi V.[4]

There are currently 1,611 Bridges students, present and alumni; 48% are first generation college students, 60% are now employed in research and development positions, 30% continue to further advance their postgraduate education. **Kelly Shepard** is in charge right now; remember her name.[5]

*One of those scientists-to-be could be you.* Are you interested? If so, look up "Bridges" or "internship" at CIRM. (2 or 3 below) It might change your life.

---

[4] https://www.cirm.ca.gov/our-funding/research-rfas/bridges
[5] http://www.guanyu3d.net/our-impact/internship-programs/index.html

# 51 But Can We Afford It?

Now we are up against it — the last and biggest question — can California *afford* to invest $5.5 billion in stem cell research, treatments and cures?[1]

Sixteen years ago as this is written, in 2004, voters said "YES!" to Proposition 71, the California Stem Cells for Research and Cures Act. From this sprang the California Institute for Regenerative Medicine (CIRM).[2]

The result? A quiet triumph.

CIRM accomplished a lot, including a systematic method of challenging chronic disease: recruiting out-of-state reviewers to evaluate proposals, in-house scientists to add their recommendations, and a large (29 members) board of directors to make the final decision, aided by input from the public.

You've read how CIRM-funded scientists and doctors saved the lives of 50 children with Severe Combined Immune-deficiency (SCID, the "Bubble Baby" disease); Our scientists restored upper body function to quadriplegics; developed two FDA-approved treatments for fatal blood cancers, published more than three thousand medical discoveries, a library of stem cell research documentation, as well as providing a measurable return of vision to several blind people — allowing one woman to see her children for the first time.[3]

This was the beginning: plowing the field, planting the seeds. New labs were built. Protocols for scientific conduct were established, in

---

[1] https://www.lbbusinessjournal.com/should-california-invest-another-5-5-billion-into-stem-cell-research/
[2] https://en.wikipedia.org/wiki/2004_California_Proposition_71#Results_of_vote
[3] https://blog.cirm.ca.gov/tag/bubble-baby-disease/page/2/

public view, in the light of day. If you wanted to participate, you were welcome. I know; I was there; at this step by-step-foundation building.

There have been 100 FDA-approved clinical trials, funded by CIRM or based on research made possible by CIRM funding.[4]

And now? If the voters decided against renewing California's stem cell research program, it would be an ending, after a great beginning.

Should we renew the program ? Some say: "We can't afford it!"

But before we throw the future away, let's consider the cost of the enemy, if allowed to continue, unopposed.

Last year, America spent roughly $3 TRILLION dollars on chronic disease.[5]

How much is that? That many dollar bills (end to end) would make a money chain long enough to reach and loop around the moon.[6]

How much is $3 trillion?

In money, three trillion is as much (in round numbers) as all of the following:

America's 2018 federal **income tax** for individuals ($1.5 trillion);

**Social Security** ($1 trillion) and **Medicare** (half a trillion) — combined.[7]

Illness healed is money saved: in the best of all possible ways. Imagine being told: "that disease which threatened your family member's life? It is now gone!"; what a joy that would be — and what a huge financial relief!

Another gigantic example: balance the cost of renewing CIRM, $5.5 billion — against one year's expense of diabetes, which last year cost America $327 billion dollars.[8] That single disease costs our country 60 times all CIRM!

Remember too, that $327 billion for diabetes is just to maintain folks in their suffering, whereas CIRM research holds a chance of making them well.

BTW: CIRM has helped support breakthroughs in fighting diabetes: one incredible therapy is a CIRM-funded ViaCyte diabetes "pacemaker".

---

[4] https://www.cirm.ca.gov/patients/power-stem-cells
[5] —https://www.cdc.gov/chronicdisease/about/costs/index.htm
[6] https://tinyurl.com/yahp4lor
[7] —https://www.stlouisfed.org/open-vault/2019/november/where-federal-revenue-comes-from-how-spent
[8] —https://www.diabetes.org/resources/statistics/cost-diabetes

Placed under the skin, it is designed to provide stem cell-derived insulin as needed, almost like a new pancreas. PEC-Direct and PEC-Encap.[9]

Here are some annual costs of four other conditions:

Cancer — $173 billion.[10]

Obesity — $147 billion[11]

Arthritis — $140 billion — [12]

Alzheimer's disease — $159 billion — [13]

Compared to such annual expense, the cost of our program is couch change.

Medical expenses are the number one cause of bankruptcy. As President Barack Obama said in his inaugural address:

"...the crushing cost of healthcare...now causes a bankruptcy in America every 30 seconds. By the end of this year, it could cause 1.5 million Americans to lose their homes."[14]

Can we afford to renew the California stem cell program?

Look deeper.

The costs of repaying the bond sales (funding source for the research) will be zero for the first five years. This was done for reasons of fairness. Those who benefit in the long run should not have to pay for everything up front.

Also, the state of California will benefit from the economic stimulus — without denying the state's ability to fund other critical programs, like education, housing, emergency services, and COVID-19.

How is that possible?

By Prop 14 law, CIRM bond sales can be no more than 0.5% (one half of one per cent) of the total bond sale capacity of California annually, leaving 99.5% available for other pressing priorities.

And — the California stem cell program has already brought in more money to the state than it actually cost. This is due to several reasons:

---

[9] https://www.cirm.ca.gov/clinical-trial/clinical-trial-directly-vascularized-islet-cell-replacement-therapy-high-risk-type-0

[10] —https://www.ncbi.nlm.nih.gov/pmc/articles/PMC3107566/

[11] —https://www.reuters.com/article/us-obesity-cost/obesity-costs-us-health-system-147-billion-study-idUSTRE56Q36020090727

[12] https://www.cdc.gov/arthritis/data_statistics/cost.htm

[13] https://www.rand.org/news/press/2013/04/03.html

[14] https://www.nytimes.com/2009/03/05/us/politics/05obama-text.html

First, when a scientist succeeds, it becomes easier for that scientist to get more grant money. For example, the Roman Reed Spinal Cord Injury Research Act of 1999 spent $15 million in California money — but the scientists did so well their work attracted roughly $87 million in matching grants from the National Institutes of Health, the Christopher and Dana Reeve Foundation and other sources — whereby a $15 million program grew to roughly $102 million.[15]

To chart CIRM'S financial effects, a study was done by Dan Wei and Adam Rose of the University of Southern California (USC)'s Schaeffer Center for Health Policy and Economics.[16]

In the measured years of 2004–2018, the study found benefits for both California and the nation.

For California, from 2004 to 2018, an economic stimulus of $10.7 billion was estimated from "gross output": "the income received by a company from its sales of goods or the provision of services."[17]

Additionally, increased revenue from state taxes ($641.3 million) was brought in from new biomed businesses and the scientists' grants.

Jobs? More than 50,000 (56,549) new Full Time Equivalent (FTE) jobs came online because of CIRM: good-paying jobs, from construction to lab technician.

California's program benefits not only our state, but the entire nation as well. In terms of economic stimulus for the United States, $4.7 billion; in federal tax revenues, $208.6 million — and 25,816 new FTE jobs.

The California Institute for Regenerative Medicine is a shining sword. We dare not let it lie in rust. We must pick it up, and fight the enemy before us, chronic diseases, threatening us all.

Can California afford to invest in a stem cell/gene therapy future? We can't afford NOT to.

---

[15] http://www.reeve.uci.edu/news-01-2019.html

[16] https://blog.cirm.ca.gov/2019/10/09/new-report-says-cirm-produces-big-economic-boost-for-california/

[17] https://corporatefinanceinstitute.com/resources/knowledge/accounting/sales-revenue/

# 52 Incurably Ill — for Just One Day?

The classic TV show *Queen for a Day* gave people a rest from their troubles. But what if a billionaire had a chronic disease — for just one day? (Wikipedia)

There was once an incredibly successful TV show called "QUEEN FOR A DAY" (1), hosted by Jack Dailey, which ran from 1945–1960.

It was an honest show in which real people told their troubles, often reducing the audience to tears. The winner was chosen by audience applause. (In later years an attempt was made to duplicate the show with professional actresses,; this was unsuccessful; apparently the audience felt the new version was rigged and turned away from it — but the original was real.)

The winner would receive substantial gifts, like a washing machine, or help with medical costs for a sick child. And it came with lots of fun stuff — being taken out to dinner in a nice restaurant, having someone else do the dishes or vacuum the rugs — a rest from chores, smiles on her face the whole day long. And every contestant walked away with at least some gifts.

Hugely popular, it is now considered the fore-runner of reality TV.

But what if it was done — backwards?

What if someone who normally had all the breaks — a billionaire — suddenly fell ill with a chronic disease?

We could call the show "INCURABLY ILL FOR JUST ONE DAY", and it would be a punishment for white-collar crime, like not paying their fair share of taxes.

The billionaire would be given a temporary form of chronic illness. But they would feel like it was forever. For 24 hours, they would be thoroughly ill.

Their money would not be allowed to buy them off. Whatever their condition, they would get regular folks' medical care: nothing special.

Sounds like Dante's definition of Hell, doesn't it? "All hope abandon, ye who enter here"... just like normal people endure, when they suffer the hell chronic disease can bring.

A friend of mine has three kinds of cancer, (lung, breast and thyroid) as well as ulcers on the inside of her throat. It is agony for her to swallow.

Some friends cannot breathe on their own, needing a machine for air.

If our fantasy billionaire wants to turn over in the bed, or get dressed to begin their day? Someone must drop what they are doing, and come and help.

Visitors? Only a few, most people having moved on with their lives.

At night, the sufferer would lie awake, immobile, waiting endless hours for the dawn. Perhaps they would escape into a dream, a beautiful vision of health — only to blink awake to grim reality.

Would I condemn anyone to a life like that? Never, it would be too cruel.

But what a great learning experience it would be for the power folks!

In a perfect world, every billionaire would experience chronic disease, for just one day…

Or maybe a year.

# 53 The Second-Most Important Matter on the Ballot

The most important matter on the 2020 ballot was of course the Presidency; but I am a lifelong Democrat, so no surprises there. My chief concern was for the second most important issue: which was, quite literally, a life and death decision.

Do I exaggerate? Let me answer that question with another:

What is the number one killer in America today? According to the National Institutes of Health (NIH):

"Chronic diseases are responsible for seven out of 10 deaths in the U.S., killing more than 1.7 million Americans each year..."[1]

1.04 million American deaths from covid, 6.5 million deaths worldwide. Each of those deaths is a tragedy: whether from COVID (total) or chronic disease (increasing every year.)

And the annual cost of chronic disease? Roughly $3 trillion dollars...[2]

Can anyone say, chronic disease is *not* a life and death concern?

But California is fighting back: with science as our weapon.

As you know, there are more than 100 clinical trials going on right now, begun or done — all of which CIRM has a connection with — from 30 projects which received early funding to 71 directly funded clinical trials.[3]

The research is of such high quality that a special federal designation, Regenerative Medicine Advanced Therapy (RMAT) was granted for seven CIRM projects. This is no empty honor. That designation is intended to

---

[1] www.ncbi.nlm.nih.gov › pmc › articles › PMC5876976
[2] https://www.cdc.gov/chronicdisease/about/costs/index.htm
[3] https://www.cirm.ca.gov/our-progress/clinical-trials-based-cirm-grants

cut red tape, and get relief faster through the FDA approval process to the patients[4]:

*What disease most interests you?* Go to the program: www.cirm.ca.gov. On your search bar, put the name of the disease, followed by "fact sheet CIRM".

For example: If you put: "Leukemia fact sheet CIRM", Google will take you to:

**https://www.cirm.ca.gov/our-progress/disease-information/leukemia-fact-sheet**

There you will find summaries of scientists' fights against this liquid cancer, their individual grants, and the total for leukemia: $193,464,356.00.

We cannot be cheap in the war against leukemia.

Unfortunately, the program's funding was almost gone.

Should we continue the fight, or quit?

November 3rd was decision day: unless, perhaps, *you voted early…*

As soon as that precious vote-by-mail envelope arrived, I ran to the kitchen table, opened the flat package, and went over the enclosed ballot, checking off every box I cared about.

I voted for President first. And then?

Proposition 14: the California Stem Cell Research, Treatments and Cures Initiative of 2020. That was a triple exclamation point issue for me!!! If I had to, I would have walked to Sacramento to vote for Prop 14.

It was not so difficult as that, fortunately. All I had to do was fill out the ballot, stick it in the envelope, seal it — and sign on the outside where I was supposed to.

Packet in hand, I ran to the post office, where "BIG BLUE" the mailbox was waiting: massive, firmly secured to the concrete: BIG BLUE, holding the hopes and dreams of all of us: who would, could, and should vote.

I stood there a moment, worried I might do something wrong.

If Proposition 14 won, the stem cell battle continued strong; if not…

I opened BIG BLUE, slid in the sealed envelope, held it just another couple seconds, reluctant to let it go. Then it occurred to me, it might look a little strange, somebody standing there with his hand in the mailbox. I released my grip, snatched my fingers back.

CLUNK, the heavy metal door swung shut.

---

[4] https://blog.cirm.ca.gov/tag/rmat/

Done.

I walked home, beaming, repressing the urge to skip.

Success on Prop 14 depends on the actions of dedicated individuals. I loved to see statements of support from CIRM board members, like Dr. Marie Duliege:

"Several of you have asked me why I am in favor of Prop 14. I'm writing this as a California scientist and researcher...

"Prop. 14 authorizes $5.5 billion to keep our stem cell research program going. This is critical: to bring this innovative science all the way to patients is challenging: requiring cutting edge research, complex manufacturing processes, regulatory approval, etc. Public money is essential because the risks are too high for a classical Venture Capital funding model. If government does not step in, very little will happen.

"The CIRM has a variety of people on its board: academics, industry executives, patient advocates... I really believe that a public agency like them can make good decisions with public money...CIRM's track record is impressive, including 2 FDA approved drugs and 64 (now 71 — dr) clinical trials.[5]

"So a lot more is coming, but it just takes time. A few examples: Phase 3 trials in ALS (Lou Gehrig's disease), Phase 2 in Retinitis Pigmentosa, etc.

"In sum, this measure will serve important unmet medical needs. "
— Anne-Marie Duliege, MD, board member, CIRM.

Time slipped by, bringing us closer and closer to the irrevocable decision.

I thought about the great Helen Keller, blind and deaf, but an unparalleled champion for research and human rights.

She said: "I long to accomplish a great and noble task. But it is my chief duty to accomplish small tasks *as if they were* great and noble."

She also said: "I cannot do everything, but still I can do something."

You should see all the great people I worked with, on the Prop 14 campaign — most of them young, scary-talented, modern-minded

---

[5] (See full list at https://www.cirm.ca.gov/clinical-trials.)

like you would not believe. Thumbs flying, they use phones like super-computers — I can barely answer my flip phone!

"But still I can do something…"

And I will.

November 3, 2020: My last weblog entry went out the morning of the election, urging voters to use their email chain one more time, and to pass along the following one-sentence message:

**"VOTE YES! on Proposition 14** — and ask a friend to do the same."

And suddenly there was no more time; the campaign for Prop 14 was over.[6]

---

[6] https://www.cirm.ca.gov/our-impact/funding-clinical-trial

# Staying Out of Politics?

"All that is necessary for the triumph of evil, is for men of good will to do nothing." — Psalm 94: 1–3, often attributed to Edmund Burke.

As you may have guessed by now, I am passionate about the California Institute for Regenerative Medicine (CIRM), the creation of Bob Klein. I believe that CIRM will drive possible cures for many diseases, saving literally millions of lives and billions of dollars: reducing misery, and lowering the cost of health.

But our beautiful stem cell program was only possible because of a political technique called the **Ballot Initiative**. This allows regular citizens like yourself or me to suggest new laws or programs in our state, for the voters to decide.

This is pure democracy; shall we do X or Y, yes or no?

But the Ballot Initiative may soon be gone.

"In 2021, Republicans have introduced 144 bills to *restrict the ballot initiative* processes in 32 states…"[1]

Hold that thought.

A scientist friend once told me: "I make it a point to stay out of politics."

That is, of course, a political position. Politics is decision-making. To not take part means just to let others control the outcome. And what if they are wrong?

Consider: one political party believes Earth is threatened by global warming, and we should do something about it. The other party says, it's a hoax, do nothing.

---

[1] https://www.nytimes.com/2021/05/22/us/politics/republican-ballot-initiatives-democrats.html

One party says the rich do not pay their fair share of taxes, and we should do something about it; the other party says, the rich should pay zero taxes, supposedly so the rich will "invest" their money in ways that drive the economy for everyone, but as we've seen over the decades since that view was first posited in the US, that's not true. They just get richer and use their money to stay that way, while everyone else ends up with smaller and smaller pieces of the pie.

One party says, vast numbers of people are denied the right to vote, and we should do something about it; the other party says —

Wait a minute; let's consider what they say — and do.

Imagine you are physically blind. Because you cannot see, a friend at home helped you fill out your ballot. But when you take that ballot to the voting booth, it is snatched from your hands, set aside, essentially tossed in the garbage can.

Why was your vote just taken away?

"You don't have a have a government-issued photo ID card."

If you are blind, you will not have a driver's license. Again, in many states that disqualifies you from voting. This affects a lot of potential voters. How many?

Twenty-one million Americans (11% of U.S. citizens) do not have a government issue photo ID (which means, almost always, a driver's license).[2]

But what about alternative forms of ID?

In Republican-controlled Texas, a gun license will entitle you to vote — of course, blind people do not do a lot of shooting, so you may not have one of those. A college student ID? That should count as valid ID, wouldn't it? Not in Texas. Their motto should be apparently, is: shooters yes, students no![3]

How about a passport as a voter ID? If you are rich and travel, of course you will have a passport. But most disabled people do not have a lot of extra money.

Oh, but the Republicans say, (and this is a 100% Republican effort, make no mistake) they offer a free voter's card. All you have to do is bring the required documentation (a birth certificate, social security

---

[2] https://www.aclu.org/other/oppose-voter-id-legislation-fact-sheet
[3] https://www.statesman.com/news/20160923/politifact-take-gun-license-but-not-student-id-to-texas-polls

card, perhaps more, varying state by state) — and pay the fee at the Department of Motor Vehicles (DMV).

Do you know where your birth certificate is? If not, obtaining one will cost you a fee, plus you will need documentation, and there will be time lost from work, putting everything together at the DMV.

And what if there is no DMV near you? In some portions of rural Texas, you may have to drive as far as 170 miles to reach the nearest DMV office. Also, some of those offices are open on very limited hours, like one day a week. Imagine your frustration, driving three hours, only to find the DMV is closed. And, of course, some cannot drive, or afford the cost, even if they are physically capable.

How about automatic voter registration, so you register once and are set for life? Republicans want to get rid of that too.

Lines of voters too long, discouraging voters? Republicans want to make them still longer, by reducing the number of days a person can vote. And who decides how many polling places there will be? If you have lots of polling places for people in a rich neighborhood, and only a few in a poor one, that will affect the results.

It boils down to this: big turnouts generally favor Democrats. Republicans, therefore, are trying to make voting difficult, especially for minorities, the disabled, students, and the poor — i.e, Denocrats.

In practical terms, what if you are a single Mom, working two jobs, and you can't afford to lose a day's pay to vote? An absentee ballot is just what you need — but Republicans are trying to take those away.

Wait, it gets worse. Those who help register voters are threatened with $5,000 fines and up to ten years jail time for even an honest mistake in a long list of complicated rules. Highly-respected non-profit groups like the League of Women Voters may no longer help register voters, because of the risks of fine or jail.

Speaking of jail, Republicans want to impose imprisonment for anyone giving a drink of water to someone in a long voting line[4]; they want to send "observers" into voting chambers, to "oversee", but in reality, I think, to intimidate those likely to vote Democrat — i.e. students, the poor, voters of color, and the disabled.

What does this have to do with the California stem cell program?

---

[4] https://www.cnn.com/2021/03/26/politics/georgia-voting-law-food-drink-ban-trnd/index.html

One Republican voter suppression goal is to **ban state initiatives,** which gave us our beautiful, voter-approved stem cell program![5]

But maybe I exaggerate? Find out for yourself.[6] And this above all:

If Bob Klein and friends had "stayed out of politics" — we would not have the California Institute for Regenerative Medicine — and all its possibilities of cure.

---

[5] https://www.nytimes.com/2021/05/22/us/politics/republican-ballot-initiatives-democrats.html

[6] https://www.washingtonpost.com/politics/interactive/2021/voting-restrictions-republicans-states/

# 55 A Child's Life

Leukemia may steal a child's life; Dr. Maria Grazia Roncarolo wants to block that theft. (www.med.stanford.edu)

Years ago, there was a boy (we'll call him "J") who had leukemia. He was about six years old, on chemotherapy, and his hair was gone. Gloria was friends with J's Mom, and they were visiting one evening.

My older sister Patty had been taken by the disease, which would later claim my younger sister Barbara as well.

This was in my early years, scuba diving at Marine World Africa USA, that most beautiful aquarium-zoo in Northern California. I loved to give slide shows at schools, talking about the joy of the sea — sharks, dolphins, eels, seals and killer whales.

Would J like to see the slide show, I asked? He nodded. I set up the projector, and began.

I always poured out everything I had in the shows. I was on a mission. I loved the ocean, and wanted the children to understand its vital importance.

But never had I worked so hard as I did that night, for an audience of one. I thought about Patty, and wanted desperately to make him happy, if only for an hour. He was in his pajamas,("He is more comfortable this way", said his Mom) looking very frail. She held him on her lap.

J appeared to like the show, laughing in the part with the giant sea turtle Chopper, who confused my hand with food. I showed J the scar. Of course he enjoyed the fight between lions and tigers (the lions won), and all the action —

"He's tired now," said his Mom. J nodded, once. He was polite about it, shook my hand and said thanks, but was plainly exhausted. We said goodbye, and they went home.

A couple months later, I heard what happened.

J and his Mother were in the living room, and she was holding him, and J said:

"I'm tired, Mom. It hurts too much. I just want to let go."

"It is all right, son," said his Mother, "If you want to let go, that's fine, I love you. Everything is all right."

She held him in her arms, until he slipped away.

Years passed, as they do. The strength and weakness of time is that it goes on, even when it seems the whole world should stop and take notice.

Just yesterday, I saw Bob Klein in the office, and told him I was going to interview a woman scientist who had been knighted by the President of Italy!

"Her name is Maria — " I blocked on the last part of her name. couldn't get the word out. (I still stutter some, even now, years after the stroke.)

"Maria Grazia Roncarolo", he finished for me, "But she is probably in Italy now."

How he knew that, I don't know, but that is typical of Bob. He likes to know what is going on, and generally does.

A few moments later, I had my phone on speaker and was talking to Professor Roncarolo — who was indeed in Italy.

She is working with a type of T cells, called regulatory T cells, which are part of the body's immune system and play a key role in maintaining its balance.

"Could they help fight leukemia?", I asked.

"That is what we are going to find out," she said.

What did she think about CIRM?

"The best thing that ever happened to the field!", she said, "And not just for stem cells, but as a model for science funding everywhere — incredible impact, helping bring cures from bench to bedside."

She had earned her medical degree at Turin, Italy, after which she transferred to Lyon, France — from there to Palo Alto — back to Italy to direct the San Raffaele Telethon Institute for Gene Therapy (TIGET) at Milan — and then to Stanford University, where she established the Center for Definitive and Curative Medicine.

But the "what" was more interesting than the "where". As I waded through materials printed out from Stanford about her, I kept shaking my head — what an amazingly helpful life. As a young physician scientist she was frustrated by the limited therapies for children with devastating genetic diseases and blood cancer. To fight against this lack of hope she decided to devote her research and clinical work to find a cure for patients with incurable diseases.

Dr. Roncarolo is a pediatric immunologist... Think what that means — to seek out flaws in a child's immune system, and try to find ways to repair them. In France, she had pioneered the effort to cure 'bubble boy disease" also known as SCID.

From her lab: "Dr. Roncarolo was a key member of the team that carried out the first stem cell transplants given before birth to treat (this) genetic disease. She also developed (a) successful gene therapy by adding a healthy gene in blood stem cells of patients with SCID due to deficiency in the adenosine deaminase enzyme. This product is now a drug approved for the European market with the name of Strimvelis, and is available for every patient in need."

Naturally I asked her if she knew Don Kohn of UCLA, who had championed that effort in the states, and she immediately responded about him in warm and generous terms, stating that they had been friends and colleagues since the beginning of their careers.

She worked to make "homeostasis" (balance) in the children's immune system. By studying children successfully cured with blood stem cells from healthy donors she discovered a "new class of T cells, called T regulatory type 1 cells, (Trt1) which help...assist the immune system in tolerating transplanted cells and organs"...(and) completed the first clinical trial using (these) cells to prevent severe graft-versus-host-disease in adult leukemia patients…".[1] These Tr1 cells can also improve the outcome of stem cell transplants in children with leukemia.

For instance, suppose a stem cell transplant from a healthy donor was needed to replace the blood cells in a child with leukemia. The body might reject the transplant — OR — the transplant itself might attack the body it was meant to help! That condition (Graft-Versus-Host Disease, GvHD) can be fatal. Tr1 cells can prevent these devastating complications and also help the donor immune system to fight the leukemia.

She studied IPEX, (actually a whole series of child diseases, which takes it name from its symptoms: Immune dysregulation, Polyendocrinopathy, Enteropathy, X-linked problems…) also sickle cell and thalassemia, and other genetic diseases I had never heard of, like the Wiscott-Aldrich Syndrome, and Metachromatic Leukodystrophy.

Dr. Roncarolo's work on cancers of the blood and genetic diseases established her as an international star in the field of stem cell therapy and regenerative medicine. When asked which accomplishment she is most proud of, she quickly responds:

"To know that there are children who have a healthy and happy life thanks to our research and determination to translate the scientific discoveries into therapies and possibly cures is incredibly rewarding!".

Let me close this chapter with another story, which for me sums up the fight.

My grand-daughter Katherine (then 6, the same age J was when he passed), and Gloria and I had driven down the coast to a lovely little amusement park, which we thought K might like. We had been there before, Gloria and I, and it was very peaceful and low key, a little lake with painted wooden swans you could pedal on, and unusual foods to eat like fried artichoke hearts (delicious) and vegetable ice cream (disgusting) — but when we got there, it was closed!

Katherine had been cooped up in the car for way too long, and the parking lot seemed empty, so we let her get out and run around for a while, stretch her legs.

And then I saw it: another car, racing from between two buildings, hurtling toward my grand-daughter.

She was a hundred feet away. I could not pull her back.

All I could do was yell, loud as I could: "KATHERINE, STOP!"

---

[1] https://profiles.stanford.edu/maria-grazia-roncarolo

She froze.

The car **whooshed** past her, so close, she could have touched the hurtling metal. I thought I saw her hair lift, in the rush of the vehicle's passage.

I do not know if the driver saw her or not. The car neither stopped nor slowed, just kept right on going.

And Katherine turned around, and skipped back to the car.

She is 12 now, and we talk a lot about her future sometimes, when I drive her to school or help her with homework.

Until just recently, she had wanted to be a marine biologist, because she loves the creatures of the sea.

But now she was considering a different path.

"I'd like to be a cardiologist," she said, "and do heart operations. They save lives, you know."

May Katherine Reed have many years to fulfill her dream.

And may Maria Grazia Roncarolo help bring an end to the vile disease which took Patty, Barbara, and J.

# 56 Disassembling the Eiffel Tower?

The glorious Eiffel Tower was originally meant to last for only a few months, and then to be destroyed. Fortunately, someone changed their mind. (Wikipedia)

On a subway to the Eiffel Tower, someone put their hand in my pocket.

It was Gloria's and my 40th anniversary, and we were celebrating it in Paris, the most beautiful city in the world.

It was our last day of the 5-day trip, and we were exhausted: physically, emotionally, financially; our credit cards whimpered — but our brains were overloaded with memories.

So much: Monet's gardens, lily ponds so lovely, I got lost and almost missed the tour bus; the Louvre museum, five football fields of art treasure including a huge painting of Napoleon crowning himself, as the Pope sits by ignored and infuriated; the Musee du Branlee, with its amazing exhibit on Tarzan of the Apes; a sidewalk café by a gold-plated statue of Joan of Arc; the shimmering Hall of Mirrors at the Palace of Versailles, and so much more...

But Gloria wanted one more present for the grandkids, and I wanted a close-up look at the incredible Eiffel Tower...

The train was jammed. Gloria got the last seat. Standing, I somehow became surrounded by four beautiful women, who seemed to know each other, although they did not speak. The smallest one of them smiled shyly at me... The jolt of the train pushed her closer to me, so that we almost touched.

Hmf, I thought, I must still have "it". I sighed for what would never be, and went back to reading the tourist booklet.

The Eiffel Tower, like the California stem cell program, began as something temporary. It was built for the 1887 World's Fair, after which it was to have been destroyed. A condition in its contract was that it must be easily disassembled.

Some Parisians hated the tower. It was physically taller than any of the churches, and for that was called anti-religion. Artists dubbed it a "tragically-designed lamp post", or "an odious column, whose shadow blots our fair city".

A campaign was launched to destroy it, backed by Alexandre Dumas, author of THE THREE MUSKETEERS . Another great writer, Guy de Maupassant, hated the tower so much, he reportedly ate lunch *inside it* every day — because, he said, it was the only place in Paris where he would not have to see the eyesore!

The tower was required to come down, to be destroyed — but the designer did not want that. Eiffel hunted for ways to show the Tower's usefulness and practicality. It was finally saved because of the new science of radio-telegraphy. The tallest building in the world turned out to be a great place for a radio transmitter.

And then — World War I.

The Germans were closing in on Paris. Relief soldiers had to be gotten to the front lines, quickly — radio signals from the Tower coordinated a wave of taxi cabs, preventing traffic jams. That single action prevented a quick victory for the enemy.

The war devastated France. Almost an entire generation of young men was wiped out in a single horrific encounter, Verdun, called the most terrible single battle of all time.

When the second World War arrived, it burst upon a still-weakened France.

The Eiffel Tower became a symbol of defiance. Their country overrun by the Nazis, resistance fighters cut the cables to the elevators, denying Adolph Hitler his moment of triumph at the top of the tower. And when the Nazis did manage to hang their odious flag from the symbol of Paris, a brave fighter climbed up the tower, and cut the hateful swastika down, replacing it with the flag of France.

I felt a tingling at the front of my left thigh, where my wallet resided.

I pivoted, saw a small hand jerk back from my pocket, disappearing up a voluminous sleeve, like a gopher returning to its hole. My billfold, luckily, had been jammed crosswise in the pocket, and was still there.

What to do? I bore them no ill will. How miserable a life must be, if the only way you could get by was to steal. Besides, I had seen on TV that if you did make a big fuss, the thieves were ready for that and would scream and yell right back, embarrassing me more than I would them.

So I called to my wife Gloria, waited till she was looking at me, then raised my hand above the crowd, pointing a finger toward each of the four.

"Remember their faces," I said.

The women glanced at each other.

The doors hissed open; they were gone.

And suddenly, there it was, the Eiffel Tower, not black as I had thought it would be, but dark brown in color, earth-tones, reaching upward, a soaring height.

Halfway up the side, a tiny moving bump. It was a person, mountain-climbing up the side of the Tower. We took his/her picture, and Gloria shopped, while I just stared, up and up at the architectural marvel.

Nervous that the wind might bend or break it, the government charged Eiffel with full responsibility. If a piece fell off and killed somebody... he would have to pay all damages. But the Tower was designed, "to takes the shape of the wind itself", as its maker said, and it merely swayed, adjusting, not warping, not breaking.

Superbly strong, but delicate, a lacework of metal; if you melted it, I am told, the tower would make a puddle no wider than the base — and just three inches deep.

But what if the Eiffel Tower had been destroyed, according to the terms of the original contract?

What a loss that would have been! In financial terms, more than two hundred million paid admissions since then, gone, and the accompanying tourist purchases.

And its value in beauty, as it lifts the soul of humanity? Immeasurable.

Like the Eiffel tower, the California stem cell program was originally built to be temporary, to last only as long as its $3 billion dollars in funding: the bond sales. When the money ran out, so would the program.

At two hundred ninety-five million a year, the life of the program would be about fourteen years: with the addition of the loan program, which meant some funds were returned to be used again, maybe seventeen.

I don't pretend to understand money. But I can recognize something wonderful, and CIRM is that.

The problems CIRM was built to fight have not gone away. Chronic disease and disability is bankrupting America. The inability to pay health care costs has been cited as the number one reason for home foreclosure.

An estimated 40% of Americans suffer chronic illness or injury, the modern equivalent of the Black Plague, which devastated Europe in the Dark Ages.

In my country alone, we lose four thousand citizens to chronic disease: the equivalent of the September eleventh massacre — every day!

Our people are dying, or suffering permanent injury, as if they were in war; shall we not defend them?

The California Institute for Regenerative Medicine is that defense: to save lives, ease suffering, and re-invigorate the economy.

To let CIRM end would be like disassembling the Eiffel Tower.

# 57 Your California College — Was It Helped by CIRM?

If I was to speak before your college's alumnae group, those public-spirited citizens who work so hard to raise funds for your school, I would say:

How much money did the California Institute for Regenerative Medicine (CIRM) provide for your scientists? The answer might surprise you.

Few people realize how many (and how much) California colleges and universities benefited from our stem cell program. For example: consider the University of California at Irvine (UCI). CIRM provided almost one hundred twenty-five million ($124,830,106.00) dollars in support of UCI projects.[1]

Partly because of CIRM funding, UCI is now a world center for neurological research, battling diseases of the nerves, including: epilepsy, Parkinson's, paralysis, multiple sclerosis, stroke, dementia, migraines, Alzheimer's, Huntington's, more — and that is just one university!

How many stem cell research dollars went to YOUR California college or institution?[2]

(For your convenience, a partial list follows later).

Another example: Stanford University: it is huge. Probably no medical institution in the world has produced more stem cell research, treatments and therapies.

Add Stanford's strength to the power of CIRM, and you have something great,[3] beginning with a building: the magnificent Lorry I. Lokey Stem Cell Research Building — how was it paid for? The great philanthropist Lokey donated $75 million, while CIRM provided $43,578,000.

---

[1] https://www.cirm.ca.gov/our-progress/institutions/university-california-irvine
[2] https://www.cirm.ca.gov/our-progress/funded-institutions
[3] https://www.cirm.ca.gov/our-progress/institutions/stanford-university

But a building is nothing without researchers; Here are just a few of the many Stanford scientists who received grants from the California stem cell agency.

Dr. Helen Blau: stimulating muscle cells to counter age-induced weakness;

Dr. Alan Cheng: working to restore inner-ear function — and hearing;

Dr. Kyle Loh: developing new sources of liver cells;

Dr. Theodore Leng: working on a product to treat blindness;

Dr. Bertha Chen: developing smooth muscle stem cells for urinary incontinence;

Dr. Michele Calos: making a therapy to overcome Duchenne disease;

Dr. Alfred Lane: treating the dreaded skin condition, Epidermolysis Bullosa.

Is such research practical, usable, real world stuff ?

Let's put it this way. One CIRM-assisted cancer remedy developed by Irv Weissman of Stanford was recently sold — to Gilead, Inc. — for **$4.9 billion** dollars — to further develop the therapy and make it available to patients.[4]

How much did California invest in Stanford scientists? $387 million.[5]

Just for fun, look up your college, university or biomedical institution, on the list below. (Note: there are many wonderful private enterprise or charitable sources of research, which are not listed here; I wanted to focus on schools for this list.

Clarification: After the name of the institution, you will see a number in **bold**; that is the number of grants which were awarded.

### Institution — number of Awards — Amount

Stanford University **138** $387,818,024
University of California, Los Angeles **114** $305,913,079
University of California, San Diego **108** $231,259,030
University of California, San Francisco **88** $198,054,592
University of California, Davis **58** $143,156,677
University of California, Irvine **50** $124,830,106
City of Hope, Beckman Research Institute **39** $116,647,699

---

[4] https://www.bloomberg.com/news/articles/2020-03-06/stanford-doctor-to-get-191-million-for-cancer-fighting-biotech
[5] https://www.cirm.ca.gov/our-progress/institutions/stanford-university

University of Southern California **37** $113,751,804
Cedars-Sinai Medical Center **24** $72,107,048
Gladstone Institutes, J. David **32** $56,437,364
University of California, Berkeley **25** $56,272,739
Salk Institute for Biological Studies **22** $53,317,589
Scripps Research Institute **23** $50,582,424
Children's Hospital of Los Angeles **14** $32,401,897
University of California, Santa Cruz **12** $26,154,919
University of California, Santa Barbara **10** $15,285,163
San Diego State University Foundation **4** $9,058,049
University of California, Merced **6** $8,724,275
Cal State Univ, San Marcos **4** $7,756,700
Cal State Univ, Channel Island **2** $7,195,995
San Jose State University **2** $7,190,853
San Francisco State University **2** $7,159,893
Cal State Univ, Northridge **2** $6,574,769
Cal State Univ, Long Beach **2** $6,496,193
Pasadena City College **2** $6,153,881
California Polytechnic State University, San Luis Obispo **2** $6,112,735
Humboldt State University **2** $5,960,993
Cal State Univ, Fullerton **2** $5,783,901
Cal State Univ, Sacramento **2** $5,432,565
Cal State Univ, San Bernadino **2** $5,421,221
University of California, Riverside **6** $5,370,973
City College Of San Francisco **2** $4,753,419
California Institute of Technology **5** $4,173,458
Palo Alto Veterans Institute for Research **4** $3,031,998
Berkeley City College **1** $2,227,230
University of California San Diego **1** $2,167,200

Imagine if all that money had been denied. But Prop 71 succeeded, and financially encouraged all those places of learning.

And if it could be continued? That is the hope of Prop 14 — **to** *regenerate* **the California Institute** — for *Regenerative Medicine*.

Talk about advanced education!

# 58 What We Have Won

Katie Sharify (shown in Paris) was one of the world's first recipients of a CIRM-supported stem cell operation, and is now working for CIRM as their Communications Team Coordinator.

When I first heard the news, I could not accept it. Surely we had done everything humanly possible, it would not be fair, if we —

I could not trust anybody else; I had to hear it from the one man who would know absolutely and for sure. I called Bob.

The first thing he said was: "Are you all right?", because it is not usual for me to call him at home.

I assured him I was fine, and then:

"I have to hear it from you, Bob, you personally, I won't believe anybody else. Did we — did we — win?" I braced myself.

"Yes," he said; "We won,", he said.

I burst into laughter and hit the kitchen table so hard I felt the fibers of the wood begin to separate, against the base of my closed fist. Another such whack and I would have needed a new table!

Bob's wife Danielle was listening in. She said later that my laugh was the most joyous thing she heard in the entire campaign. I must concur.

That moment? One of the greatest moments of my life. When I knew we had won -against all odds — $5.5 billion for stem cell research? It was like the birth of my daughter Desiree; or seeing Gloria in her white lace wedding dress; or Roman's leaping backhand catch, full-stretch high off the ground, to win a baseball championship — it is not safe to feel such joy! One might explode.

"Thank you", I said, and Bob started to make the usual noises, about how it was a team effort and he could never have done it without me and all the patient advocate community, turning the compliment back on the giver, like he always —

I wasn't having it. Not today, not this time.

"No," I said; "Everybody helped, but you made it happen. You were the irreplaceable man. Thank *you.*"

And just this once he let it slide, acknowledging it, accepting the compliment.

"You're welcome," he said.

I hung up the phone, sat down abruptly, almost missed the chair, burst into tears —

And then I called my son.

It was a very near thing. We started out with a strong lead, nearly 20 points ahead, but COVID frightened off many voters.

In the end we had eight million, five hundred eighty-eight thousand, six hundred eighteen "YES" votes. (8,588,618)

The opposition got eight million, two hundred twenty-two thousand, one hundred fifty-four "NO" votes.(8,222,154)

8,588,618 minus 8,222,154 equals 366, 464 — Our margin of victory? three hundred sixty-six thousand, four hundred sixty four votes.

Percentage-wise, we had 51.09%; they had 48.9%: it was a 51/49 victory.[1]

Against all odds...

What had we won? What will be the impact of $5.5 billion dollars in regenerative medicine, on top of the $3 billion early funds, from Prop 71?

It will take generations to fully answer that question. Unique in all the world, CIRM will forever deserve study as the place and time when a state said "Enough! No More!", refusing to accept chronic disease and disability.

CIRM's renewal meant positive change in areas where hope had been slim to none. Want some examples? Here are fifteen, all opinions:

1. Medicine: the field of healing is no longer just a way to ease symptoms. With regenerative medicine, the goal is cure, to make people *well*. Not better, well.
2. Aging: instead of a fixed and inescapable downward slide to death, aging may now be an illness to be fought, victory bringing long and healthy extra years.
3. Pain relief: instead of drugs to mask a problem, regenerative medicine may heal a genetic error, or regrow/replace damaged nerves or tissues.
4. Leading by example: after Prop 71 succeeded, other states began stem cell programs: Maryland, Minnesota, Washington, New York, Massachusetts, Connecticut, New Jersey, Missouri and Michigan. Only Texas had similar funding ($6 billion), although it came with restrictions, such as no embryonic stem cell research. New Jersey attempted a $450 million program, but their campaign lost narrowly, opposed by the billionaire Koch brothers' organization, Americans for Prosperity.[2] It is to be hoped that these and other state efforts will be funded.
5. Political science: how were patient advocates of widely differing diseases able to unite, and convince a state to spend literally billions of dollars on medical research? Bob Klein and the team at the Americans for Cures Foundation are assembling an extensive archive at Stanford: a systematic look at how the thing was done.

---

[1] https://ballotpedia.org/California_Proposition_14,_Stem_Cell_Research_Institute_Bond_Initiative_(2020)

[2] https://www.ncbi.nlm.nih.gov/pmc/articles/PMC5812675/

6. Economic benefits: money spent on developing cures for chronic disease does not subtract from the economy but adds to it: bringing economic growth to the state and nation, while hopefully helping to reduce medical bankruptcies.

7. Health: no luxury compares to the relief from pain, nor the glow of regained good health. Regenerative medicine should be a continuing priority.

8. Ethics: in addition to carefully prescribed patients' rights, freedom to research must be protected against the attacks of religious/political ideologues.

9. Public involvement in decision-making: CIRM's final grant decisions are made with input from the public. Anyone who wishes may get involved.

10. Mental Illness: CIRM treats mental illness as sickness with a biological component: a medical problem to be solved, not a shame to be covered up.

11. Payment-reimbursement model: Instead of the current make-payments-till-the-grave standard, we need a model for reimbursement. Regenerative medicine offers the possibility of a single-dose therapy — how much is fair?

12. Affordability: the renewed CIRM will have a new 17-member committee to study ways to make therapies affordable and accessible: first for those in the trials, and from that, hopefully, for the general public.

13. Transparency: No other government agency has a permanent oversight body, the Citizens Financial Accountability Oversight Committee (CFAOC). As always, CIRM will undergo multiple audits and reviews, including a National Academy of Sciences study to update our Conflict of Interest policies.

14. Public-private partnership: CIRM can make interest-bearing loans to scientific companies, with increased return on California's investment.

15. Democracy in Action: CIRM was voted into law by direct democracy: a citizens' initiative. As Robert Klein (the younger) put it :

"The people spoke; we were heard — and we prevailed."

(**Below is a chart of the campaign**, in two versions; please use whichever works best.)

PDF FILE OF THE CAMPAIGN

file:///C:/Users/Don.Reed/Downloads/2020%2012%2010_CA%20for%20
Stem%20Cell%20Organization%20Chart.pdf
ALTERNATE FILE OF CAMPAIGN (use whichever one copies best)
file:///C:/Users/Don.Reed/Downloads/2020%2012%2010_CA%20for%20
Stem%20Cell%20Organization%20Chart(11)%20(1).pdf

# 59 Unfinished Business

CIRM is still a work in progress; ask Geoff Lomax about the Community Centers of Excellence. (www.blog.cirm.gov)

"Though much is taken, much abides; and though
We are not now that strength which in old days
Moved earth and heaven — that which we are, we are;
One equal temper of heroic hearts,
Made weak by time and fate, but strong in will
To strive, to seek, to find, and not to yield." — "Ulysses", Alfred Lord Tennyson

Remember Professor Maria Grazia Roncarolo of Stanford, previously knighted by two Presidents of Italy? On September 23, 2021, she won a CIRM grant for nearly $12 million ($11,996,634) to do a clinical trial on T cells — to fight leukemia, that vile disease which killed my sisters, and a little boy called J...

So, is this the end? At 76, I am admittedly slowing down a little, but still have many goals, personal and professional: places which need pushing. Some chores leap out: issues which demand support. (Others will suggest themselves, I am sure, as time goes by.) These include:

1. If the fates (and the publishers!) allow, I would like to write another stem cell book: this one on CIRM's greatest challenge: the $1.5 billion dollar set-aside for neurological conditions. It will be timed for the 20th anniversary of the California stem cell program.

2. I must work to protect voting rights — especially citizen initiatives like the one which built CIRM.

3. The right to research must be protected. If ideologues take power, some of our most important forms of research may be banned.

4. We need a cure for paralyzed people — including, of course, my son. I want to see him on his feet before I die.

5. Pancreatic and all other cancers should be wiped off the earth. This will be an appropriate compensation for the loss of my beloved Gloria.

6. Cure research must be understood as a stimulus to the economy, to make more jobs, and to lessen medical bankruptcy: not a take-away, but an addition.

7. Repairing blindness should be made routine: make an appointment, have minor surgery, insert stem cells, go home — and see again.

8. Aging should be viewed not as an all-conquering inevitability, but rather as a condition against which to struggle, and with more victories than earlier generations enjoyed. I hope to take part in an anti-aging clinical trial — and live as long as my father, Dr. Charles H. Reed, who just celebrated his 100th birthday!

9. Mental illnesses should be healed: a sickness with a biological component.

10. Other states should be offered assistance in developing their own regenerative research programs. International cooperation should be fostered, patterned on what I think of as Bob Klein's "California method". What worked in the Golden State might be useful across the planet.

11. I must take part in some way in the environmental protection struggle, which is too big for anyone to ignore.

12. A two-part piece of the puzzle must be put in place: the Alpha Stem Cell Clinic Network, and the Community Care Centers of Excellence.

It might be simpler to call it the "Alpha Trials Network", which has at least the benefit of fewer words. Bear with me; this is important, and wonderful.

Plans are still works in progress, not ready for public viewing.

As Geoff Lomax, (Project Manager, Government Affairs) put it: "The goal is embodied in Prop 14. But CIRM has not proposed anything formally and the ICOC has not made a decision yet."

That noted, here is my small understanding.

The goal is to *expand* the Alpha Stem Cell Clinic Network, and *establish* the Community Care Centers of Excellence, at which **clinical trials** are conducted — and treatments made available to the patients.[1]

The "Alpha" program (ASC) has been in operation since 2015. With six top-notch medical centers — City of Hope, UC Davis, UC Irvine, UC Los Angeles, UC San Diego and UC San Francisco — The ASC has sponsored "more than 100 clinical trials, treated 750 patients, and brought in more than $57 million in contracts with commercial sponsors".[2]

The Community Care Centers will increase the capacity of the program to conduct clinical trials and "make the resulting treatments and cures broadly available to California patients." (1. Ibid)

These headquarters for testing, will be a reliable alternative to "rogue" establishments, which often peddle non-FDA-approved "therapies". Instead, patients will take part in carefully designed human trials, safe as humanly possible.

Researchers must include a plan to enhance patient access to the trials, regardless of geographic location, or financial status.[3]

Researchers may sift through information on 20 million potential clinical trial volunteers.[4]

One planned empowerment is the synergy (increased strength through cooperation) between the various sites. As CIRM President Maria Millan puts it: "We can share data and information...to improve

---

[1] Section 125 290.72 addition to the California Health and Safety Code
[2] "A Patient-Centered Cell & Gene Therapy Clinical Trial Platform", CIRM Alpha Clinics
[3] http://clinic.stemcell.uci.edu/About/history.php
[4] https://www.cirm.ca.gov/patients/alpha-clinics-network/alpha-clinics-trials

the quality of both the research… and the clinical care we…offer to the patients."[5]

This is where the rubber meets the road: where testing will be done, to see if new therapies can reduce pain, restore strength, enhance health, and save lives — which is what the California stem cell program is all about.

With all the political fighting nowadays, is our program safe? CIRM itself is legally well-armored, hopefully safe from being overturned.

There are no guarantees, of course. Attacks can be subtle and long-ranging; even if they do not hurt CIRM directly, they may prevent other states from developing similar programs.

But those are battles for another day. We have won a great victory: Bob Klein and the patient advocates of California. We can and will fight on.

No matter what politics surround us, we all have families we love and will defend. Let that be our uniting factor, our hope, our shining star to follow: As we continue what VA researcher Sam Maddox[6] once called:

The Quest for Cure.

---

[5] http//www.cirm.ca.gov/about-cirm/newsroom/press-releases/12132019/cirm%E2%80%99s-alpha-stem-cell-clinics-given-high-profile-role
[6] https://www.amazon.com/Quest-Cure-Restoring-Function-Spinal/dp/0929819039

# 60 A Christmas Poem

"Let no one doubt the power of a dedicated few",
  Helen Keller meant patient advocates; she was talking about you.
  2020 was a terrible year, from January to December,
  But there was one earthquake for good, which we will all remember;
  Proposition 14 was a gentle war, in which nobody died,
  And millions gained a new lease on life, now that is cause for pride.
  Fiona Hutton led one army, metaphorically,
  Her veteran's experience was always plain to see.
  Dagny Ellenberg, Evan Swerdfeger, and Amanda Bobbit too
  built coalitions into a group, uniting folks like me and you.
  Sierra Layton and Melanie Tuberman got us publicity free,
  Their labors in earned media, put us in visible territory;
  Jon Koriel was content manager, Sarah Melbostad Press Secretary,
  Communications Director Kendall Klinger, her sights set firm on
victory.
  Paul Mandabach was another general, a warrior to respect.
  He helped to lead Prop 71 and 14, with many energies to direct.
  Brett Noble was more peaceful, but he always got things done,
  He gave a clear example of how a meeting should be run.
  There are people who are champions, though they work behind the
scenes,
  Like the person who keeps Bob Klein organized, that's Elizabeth
Tafeen;
  Linda, Lisa, Mimi, Aretha, Dana and a dozen more,
  They keep Klein Financial going, they make possible the war.
  When Bob donates millions, that is something he could not do —
  Without the efforts of Judi and Frank, Chad, Lisa and Mitu;
  Our lives were enriched by Maati, Chenelle, Sally, Dan and Annette.

Alyssa and Lauren, Bob's daughters two, the most creative minds you could get.

When next you go out to Bob's house, at one of the parties he throws,

If you want financial conversation, stand by Alan Bogomilsky — he knows.

For Danielle Guttmann's Initiatives, developing the LifeGuard invention with zeal, talk to Alicia, Amanda, Peter or Christine — they will make it real.

Danielle Guttman-Klein is not "just" Bob's wife, not only his lady fair,

Her contributions to the struggle were great, she was a co-Chair;

And now we come to the stem cell folk, each an expert in their field;

and the weapon they have in common? Information is a sword to wield.

David Serrano-Sewell has a wealth of experience,

He once served on CIRM's board of directors, knowledge you can sense.

Senator Art Torres, brought Sacramento to us,

"Gavin Newsom? Sure, I will talk to him", without making any fuss.

And one special thing I must thank him for, which he did all the while,

When Bob Klein was exhausted, Art would find a way to make him smile.

**Rob** Klein, Bob's son, is Campaign President, head of Governmental Affairs,

But you would never know it, he doesn't put on any airs;

You can bring him any problem, he will make time for you,

He understands the real estate planet and regenerative medicine too.

Mitra Hooshmand knows all about science, and can tell you in small words,

But her working ethic is "I could do that" — a volunteer like you never heard.

Anna Maybach, director of campaign communication,

She can also edit an essay, make it smooth as on vacation.

Fera Yildizoglu was very modern with Social Media,

As Consultant Advisor she was an electronic encyclopedia.

Jacqueline Hantgan, Director of Outreach, with an "F-U 2" attitude,

And by that I mean "Follow up, Follow Up" — without ever being rude.

Executive Director Melissa King, had Prop 71 in her head,

Vast knowledge working for CIRM itself, stood her in good stead.
Bob Klein... like a football linebacker, he is everywhere on the field,
In him there is no inch of quit, no particle of yield;
I have written three books on this struggle, this will be the fourth,
And yet I know I will never find words to properly measure his worth.
Now one final phrase, and then, I promise I am through,
To everyone who built this dream — thanks for the privilege of writing
about you.

# 61 Statement from Maria T. Millan, President/ CEO California Institute of Regenerative Medicine

Senator Art Torres (ret.), Vice Chair — one of CIRM's longest serving and most cheerful champions. (www.law.ucdavis.edu)

Jonathan Thomas, Chairman of the Board, had the unenviable task of following Bob Klein, but eventually blazed his own unique trail. (www.blog.cirm.ca.gov)

Maria Millan currently serves as CEO/President of CIRM — read what she says about patients' wristbands. (www.blog.cirm.ca.gov)

"On my desk is a small box containing hospital wristbands. Every one was from a person I operated on, in my days as a surgeon. I keep them to remind me how those patients (and their families) went through rough times, almost unfathomably negative events and potential tragedy. It is to ease such human suffering that CIRM exists.

For we at CIRM, Proposition 14 meant the difference between continuing our program, and watching it end. It was a time of enormous uncertainty; would California decide in our favor, to renew the program's funding, or not?

When California said "Yes!" I was overwhelmed. In this most challenging of times, we had prevailed! It was surreal. But it was true. Through all the noise and tragedy, the human capacity for resilience and hope had prevailed. I was in awe of the Prop 14 team and patient advocates who made this a reality. The passage of Prop 14 felt like a hug of encouragement — from Californians to Californians — within the research-supporting ecosystem that CIRM has made.

The CIRM team has weathered the storm of uncertainty. Now we are poised and ready to go — I feel an energy surging through the organization and our public stakeholders — it is almost indescribable. I can only say:

Thank you, California!"

# 62 An Interview with Bob Klein

Bob Klein spreading the good word about the California stem cell program — look over his right shoulder... (Americans for Cures Foundation photo)

The last word on CIRM should come from the man who began it.

Bob Klein wrote the original bill, Proposition 71, as well as the renewal initiative, Prop14, with the help of James Harrison, as well as advice from dozens of top scientists, institutional leaders and policy-makers.

He organized both campaigns, first Prop 71, gathering thirty-plus million dollars for it; he, along with John and Ann Doerr, were the largest donors, contributing roughly four million apiece. Similar amounts had to be raised for Proposition 14. Both efforts cost time from his company, Klein Financial.

When Proposition 71 passed, Bob led the program through its formative period — more than 6 years — with no salary from the Institute. Only in one six-month period did he accept compensation from the state agency Board, and that was only to cover the cost of his administrative staff, after he had announced his Board retirement in 2011, and the Governing Board of Directors was seeking his replacement,

Do you remember I once told Bob he was a very poor businessman? and he said "Wha-at?!!", and I reminded him how much money he had "lost", supporting stem cell research over the years — which never brought him a nickel. I intended the statement as half joke, half compliment, but I don't think it came across that way...

Bob responded: "My youngest son Jordan's life was at risk; and, ultimately, the quality of all my children's lives and the long term health of their children will all be remarkably improved by the therapy revolution built around stem cells and genetics. Not a dollar was 'lost'.

He paused, then said:

"That was the best gift I could give my children; and which you — Don — have given through your work on the Initiative. The work of thousands of volunteers and the votes of 8.5 million California voters have passed forward the greatest possible gift to all our children's health, to the reduction of human suffering from chronic disease, and to the health of future generations."

Other states were watching.

After our first success, six other states funded programs for regenerative medicine. Now, those state programs were mostly out of money. Could we help in some way?

There were also Federal agencies and programs to be defended, like the National Institutes of Health (NIH) and the Armed Forces Institute of Regenerative Medicine (AFIRM). These needed not only financial support, but also protection of research freedoms. There are those — primarily in one political party — who would cheerfully shut us down. That cannot be allowed.

As California goes, so goes the nation.

And now, here is an exclusive interview with the man who knows stem cell research funding better than anyone. Special thanks are due to Bob's personal assistant, Elizabeth Tafeen, for taking notes. Bob has a lot to say, and his words are not always easy to follow — but it is overwhelmingly worth the effort so to do.

Bob is "B", I am "D"

D: First, Bob, thank you for taking on this incredible challenge, Prop 14 — which I regard as the most important medical research battle of all time. It would have been easy for you to rest on your laurels after passing Prop 71, but you did not. Why did you feel the new effort was needed? Why now? And why such a staggering amount of money?

B: We were at a point in 2019 where it was absolutely clear there were therapies advancing, due to the original $3 billion of Prop 71 funding; we were saving lives of adults with cancer, and of children born with no immune systems.

With 3,000 peer reviewed, published medical discoveries, and multiple "Breakthrough Designations" for therapies that the FDA had awarded (stem cell therapies derived from California's funding), it was clear we were on the verge of a new wave of therapies, within the reach of patients and doctors.

My youngest son had died, as well as my mother who had Alzheimer's. They had been my original impetus to drive a broad initiative for new therapies in the stem cell field. Even so, it was clear — through many families and individuals I have met — that it would be immoral not to seize this opportunity: to continue California's leadership with therapy development and clinical trials, to save or improve all these lives.

You asked why this large amount of money? Well, $5.5 billion is a large absolute dollar amount; but in terms of medical discoveries, large pharma companies might commit that amount to the development of a single therapy!

With the California agency and the financial leverage it gets from philanthropists, institutions, universities, patient advocacy groups and the private sector, the first initiative's $3 billion gained $11 billion in matching funds for CIRM research, especially in clinical trials. Hopefully the $5 billion is enough to achieve similar critical mass, bringing in billions of additional funding.

Over a long enough period, the funding will be distributed to attract and hopefully result in similar leverage, to put the matching funds near $20 billion for the second initiative. In the universe of medical research and clinical trial funding, that is not a huge amount, given that California is the leader globally.

The $5.5 billion was the appropriate amount, because it had enough critical mass to sustain a broad network of research and clinical trials across the state through our chronic disease program, for the early disease intervention with stem cell and genetic therapies.

With this amount, California can hopefully serve as a surrogate *nation state:* with enough scale that it could conduct not only a breadth of research and clinical trials, but it can also specifically fund clinical trials with embryonic stem cell therapies and fetal derived stem cell therapies: to treat genetic diseases that can leave people blind, for example.

Individuals suffering from either: retinitis pigmentosa, a genetic form of eye disease hitting people as early as their 20s: or age-related macular degeneration that impacts seniors, with up to 50% of individuals between 65–85 going partially blind. These afflicted individuals may benefit from fetal or embryonically derived stem cell therapies, now in FDA approved human trials.

California is the only governmental entity in the US funding clinical trials, at scale, with these stem cell therapy origins. The National Institutes of Health (NIH), for religious issues, has not, to this day, funded clinical trials derived from these cell types: even though in both cases, the materials used would otherwise have been discarded.

D: Why did you feel the effort had to be a state initiative rather than federal?

B: First, it is critical that California provides a sanctuary for science and medicine where there is a suitable level of funding for fetal and embryonic stem cell clinical trials. As I mentioned, because of religious pressures, the NIH has not funded (as of 2021) clinical trials derived from either of these cell types, despite a growing body of proof they could make radical improvements to the lives of paralyzed people, and may restore sight to the aged, suffering from macular degeneration.

Secondly, California's funding gets tremendous leverage: sometimes 5 to 1, potentially even more, due to contributions from biotech and patient advocacy groups' donations to trials originally funded by the California agency. That leverage comes from philanthropic organizations like JDRF where I once sat on the board, or the ALS Golden West Chapter for Western US, or the Foundation Fighting Blindness, and numerous others. Such organizations recognize the value of continuity of funding; California's funds are there from early discovery through development stages to FDA clinical trials; accordingly, they are prepared to come forward with substantive matching funds.

Biotech will not even enter a new therapy development area, if they don't see a reliable continuity of funding. Fortunately, the state of California, through long term funding from bond sales, can show that for 10, 12 or 14 years (as was the case with Prop 71), there would be a long-term funding stream.

A state initiative can move therapies forward through a long-term process.

With the federal government, however, you may have a change in the congress every 2 years, and every 4 years from a presidential election. You do not know what the political balance will be; you may have a discontinuity in funding, a stop-and-start movement. That inconsistent dedication to the advancement of medical science can leave promising therapies stranded. Young scientists — or veterans who may have committed 20–30 years of their life to a therapy — may see that on the verge of success, they must leave the entire field. What else can they do, if they cannot support their family any longer, because the funding is not there?

Continuity of funding assures that the best talent can be pulled together; so FDA experts and clinical trial doctors can work together for years, with the initial discovery scientist. They must all know the money will be there, to carry them forward, so they can participate in these historic therapy developments and clinical trials.

D: Not counting COVID, as you were going into the campaign, did you think Prop 14 would be more or less difficult than Prop 71?

B: Prop 14 had some new challenges. In the 16 years since Prop 71, the numbers of public media science journalists have been decimated. Over 90% of them are gone now, replaced by sports journalists, general commentators, and social media. Also, the public attention span has become shorter, and very visual; the group that will read long science stories? That too has shrunk. This creates a huge gap between where people depend upon public communication for voting; and science that is dependent upon public support.

What is the US going to do as an advanced, industrial society that depends on science to create high quality jobs and economic development — without public media, scientific journalists? Who will explain to the public the areas of promise, advancement, and what needs funding to continue? How will we gain and maintain support for revolutionary therapies to save family members' lives, or to mitigate chronic disease through stem cell/gene therapies?

Without public media, science journalists, we must try to communicate complicated issues through social media, texting — and 10 second ads!

Another challenge was that in 2020 there were other California ballot measures that were extremely expensive, with over $1 billion spent on the Fall 2020 California Initiatives; that drove up the overall cost of ads. We had to rely heavily on the efforts of patient advocacy and civic organizations to convey complex messages, about the promise in this area, the progress, the lives saved, and the medical advances. These are not 10 second messages.

Fortunately, we had 100 patient advocacy organizations, medical societies, and chambers of commerce supporting our effort. As you know, Don, we also had the University of California Board of Regents making a precedent-setting move, to break with tradition and endorse the initiative *for the entire University of California system*. Every major metropolitan chamber of commerce in California, except one (Orange County) endorsed our effort.

From the Juvenile Diabetes Research Foundation to the Foundation Fighting Blindness, the ALS group, and various Sickle Cell foundations, the spectrum of disease advocacy groups was broad; we needed every one, and not just to endorse, but to work: gathering signatures, spreading the word. We had to build an energetic and involved coalition that would provide alternative messaging channels; that was our biggest challenge in 2020.

D: With the COVID-19 challenge, how did you adjust the signature gathering?

B: We had to understand early that Covid 19 was a broad-based threat. COVID-19 was a broad-based threat to the health of American citizens and there would soon be a shutdown across the country. So, in February, we told our signature gatherers we had to accelerate the program. We had to recruit more signature gatherers; and we did this just in time.

We got almost to our total number of needed signatures before the shutdown came. Even so, with over 900,000 signatures in hand — as of the shutdown in public signature gathering — we almost failed to get enough signatures, including the 10% additional signatures we needed as a margin — despite the fact we had reached a pace of 100,000 signatures a week before the shutdown. Thankfully we were able to continue with the advocacy groups through slow direct signature gathering and earn that important buffer of extra signatures, more than required.

D: Would you like to add anything about the campaign itself, after the signatures were gotten?

B: It is amazing what a small dedicated group of people can do, backed up by the support of a large community of grassroots and civic organizations. As you know, I was fortunate that my son Robert could serve as campaign President, coordinating all of the various consultants, along side the work of Melissa King, Jacqueline Hantgan, and you Don, Senator Art Torres and certainly Anna Maybach, who had a lot of great inspired ideas. The scientific coordination along with scientific patient advocacy educators, was executed through Mitra Hooshmand, Ph.D, who was also teaching an introductory stem cell course, at UCLA.

Think of this small group of people coordinating the various consulting groups, plus patient advocates, civic organizations and the university systems. It was inspiring to see citizen action so highly effective and passionately executed, in such incredibly difficult circumstances. We were able to reach the public with enough information that the California citizens voted YES! — in 8.5 million votes, a historic vote total for a bond initiative.

D: I always felt CIRM was perfect in every way — but you wanted changes — what were they, and why?

B: When I was working on drafting the new initiative, with the help of my son Robert, and attorney James Harrison, we brought in information from across the state: from universities, nonprofits, patient groups, individuals, and board members. We were missing a vital constituency: nurses, who are so absolutely critical to the operation of hospitals, clinics and clinical trials.

Clinical trial nurses contribute a huge amount of crucial information. Also, they are the vital link for the consistency in how patients are treated in trials. There are always written instructions, oral instructions, even video instructions, but it is the nurses who provide quality and consistency in these trials, so vital to their success. We needed 2 clinical trial nurses with specific characteristics, as I described in the initiative.

It was clear that academic, biotech and civic leadership in the advanced technologies were concentrated along the coast. Large portions of our population, however, reside in the Inland Empire, in the Riverside San Bernardino area and in the northern valleys of San Joaquin or Sacramento. The latter was represented because of UC Davis, but the San Joaquin valley and the Inland Empire were not consistently represented on the Governing Board. Accordingly, two permanent board positions from these regions were established, to provide equitable representation.

D: Would you talk about the permanent Governing Board "Working Group" on Affordability and Access?

B: We needed to make sure that not just the highly informed or those with very expensive medical advisors would benefit from clinical trials. We needed to establish real medical equity of access. So I set up a permanent "Working Group", with patient advocates, patient navigators, health care reimbursement experts, and board members.

It is part of the state funding agency (not only) to initiate programs for access but also to husband and protect the funding sources I set up to help patients participate in clinical trials and to access FDA-approved new therapies. Patients might need financial help with transportation costs to get to the clinical trials or new therapies centers, plus the cost of meals and lodging for them and a caregiver if they needed one. The moral purpose of a healthcare program requires structural elements and an appropriate level of staffing, to achieve fair access.

I was made aware of the tremendous success of the Cancer Centers of Excellence that had been established years ago, by the American Cancer Society, and the American Association for Cancer Research.

Accordingly, we put into the initiative a provision to fund community care centers of excellence. These would be particularly useful in the outlying urban centers in the central valleys, northern California, and the inland empire. Centers of excellence could collaborate with the great medical centers so a patient could get access to clinical trials, not just in the core centers of academic excellence but also in major inland population centers, all across California. This "Working Group" would focus on the functional capacity of those centers, and on the financial access to FDA approved therapies, including subsidies.

No one wanting to participate in a clinical trial should be denied by their inability to pay. We must and will find a way to include qualified participants, regardless of their financial capacities to pay the cost associated with effective access.

D: Reading the initiative, I notice additional procedures on conflict of interest. Seems like we have done so much there. Do we need still more regulations?

B: The record of the California agency has been tremendous in creating high standards to avoid conflicts of interest. CIRM has operated for 16 years with a clean bill of health. Wherever there has been an issue raised, the reports have been clear: conflict was avoided because of fire walls previously set up to protect the research. We have cooperated with the California legislature to establish policies to avoid even the appearance of conflicts of interest.

Even so, it is always important to have self-examination. The new initiative requires the agency to go back every 4 years and look at what the National Academy of Sciences is doing, and study their most advanced guidelines. We want to make sure we meet or exceed the highest standards of the field.

D: Please talk about the $1.5 billion set aside for neurological conditions.

B: The brain and central nervous system are some of the most complicated parts of the human body: extraordinary in design and function, and very difficult to understand. Working with the brain and the central nervous system requires often much more time then working with other organ groups in the body. Accordingly, I wanted to set aside a floor of funding, at least $1.5 billion of the $5.5 billion as a guaranteed allocation for this area of research.

Naturally we applaud the fast-moving new therapies dealing with other areas of chronic diseases, such as diabetes, liver, kidney, pancreas or heart disease. But we did not want all the funds to be drawn away by those areas, leaving the brain, brain stem, and central nervous system without sufficient funds for major advances in research and clinical trials.

For example, the connection between the eye and brain is critical to understanding vision problems. Same thing with hearing. With therapies being demonstrated that could restore functional sight for Age Related Macular Degeneration, individuals who are functionally blind, we are on a frontier in California where we really need to understand the topic, the treatment for the eye, but also that messaging to the brain by optic nerve. The conditions are unexpectedly complex. You can go all the way from eye diseases and hearing loss to Alzheimer's to schizophrenia to paralysis — we have a great spectrum of critical therapies and treatments that we want to make sure advance, and not be left behind, even while we applaud the fast-moving new therapies dealing with other parts of the body, with chronic disease, such as diabetes and heart disease.

D: after Prop 71, other states, including Texas, attempted to duplicate California's success. Care to comment?

B: Texas was successful in passing a citizens' initiative in about 2006 for $3 billion, copying California. Unfortunately, it restricted the funding to cancer research, largely because of the ideological sensitivity to religious arguments. Early in 2020, Texas voters passed another $3 billion; fortunately for them it was before the pandemic really shut down the communications. Without the religious issues being raised by people with ideological concerns, it was a much easier task for them to pass a new initiative. We applaud their work; fighting cancer is a vital research endeavor.

Interestingly, the first major cancer stem cell breakthrough to get to phase 3 trials and be acquired in the cancer area as a new therapy, dealing with cancer stem cells, happened because of California's support for anti-CD 47 research. This treatment succeeded because a CIRM-sponsored therapy recognized that cancer stem cells hide from the immune system, hijacking molecule CD 47, so the immune system cannot see that these are rogue cells that need to be destroyed. This is research that led to CD 47 and it was was acquired by Gilead for disastrous cancers and has a fast track designation for FDA approval. It is in phase 3 trials. Gilead paid $4.9 billion for the acquisition rights.

This therapy benefited from tremendous leverage. California invested roughly $40 million and there was a total of approximately $1 billion total invested because of leverage on California's investment. So California has had a huge impact on cancer, ahead of Texas in leading the field. But we applaud Texas and everything they are doing; they have phenomenal scientists there, and imitation is one of the greatest compliments you can give to someone.

D: Would you have any recommendations for?

B: The federal government, through the National Institutes of Health (NIH) has delivered tremendous research. But there are problems with the funding mechanism and how peer review and scientific panels are conducted.

The California structure has many more protections against bias. In the California structure, first of all, our peer reviewers are from out of state. Reviewers of a California grant cannot be affiliated with a California research institution, nor be funded by a California University. You have to be independent, either from out of state or out of the country. You cannot have published within several years with the person you are evaluating; you cannot have been their mentor when they got their Ph.D. or been involved with their graduate work. You cannot have any shared financial interests with them. All of those provisions are great, shared in significant part by the NIH; however, the NIH is missing some of the safeguards to bias and relationships.

In addition, the people that oversee you on the panel at the NIH, based on my understanding, can be individuals from your own group of institutions. So that is an affiliation that may create a bias.

We need to study how the NIH awards funds; more patient advocates are needed on the board to create more independence, and we need strong barriers to affiliation and relationships. This will help drive more innovation in the grants and provide more focus on the objective quality of the grant, rather than personal relationships biasing the outcome. We also, of course, need more funding nationally. We are starving the young scientists in this country for funds. We do not have enough programs funding them and the cutting edge innovations we need.

D: do you feel an international stem cell research/gene therapy program might be successful and would you be willing to advise on such an effort?

B: An international stem cell/gene therapy program would, I think, really leverage the capacity of the world. Such a program would represent respect for patients with chronic disease, who do not have days to lose, much less weeks or months or years. They need the synergy and power of Earth's best and brightest scientists.

That is why, as Chairman, I established an international program that brought together about 10 nations in collaborative efforts. On the first disease team grants by the California agency we allowed foreign countries and their teams to compete and join our teams, as long as the foreign nations would fund their scientists. We were able to bring 2 teams of Canadians together with 2 teams from California. which were able to lead the discovery of new cancer therapies.

The first collaborative California and Canadian joint team led to 2 trials. The other team from California and Canada was able to lead to cancer discoveries that led to 3 human trials and an early diagnosis of a deadly form of cancer. Within just 4 years, both sets of teams had clinical trials approved. This was outstanding success, in a very fast time period, especially since each of the California teams got only $20 million to work with and each of the Canadian teams got less than that from their government. A little bit of money, and a small staff, but it created a synergy of knowledge between 2 countries, California (considered as a nation state) and Canada, in a joint effort to get therapies to patients faster. Such international cooperation should be continued, and expanded.

D: Do you have any other comments you would like to make about the California stem cell program?

B: The vision of the California voters was inspiring. It is a huge amount of effort to write and run an initiative campaign. As I said before, fortunately, in Prop 14, my son Robert, President of the campaign, was able to run the initiative with a small staff of very dedicated people.

That effort took place in the middle of a pandemic, when the political future of the country was also at risk, pulling people's attention to the very important politics of the moment; but still the California voters gave us enough mindshare to focus on the complicated issue of a stem cell and genetic research and clinical trial infrastructure for the future of healthcare in the 21st century. They trusted the California program with an additional $5.5 billion at a time when they were worried about the economy.

This was the largest vote in the history of the state in support of a bond measure.

Think what vision this was, on the part of the California voters; what a commitment to early intervention breakthroughs for the 21st century. They recognized this was not just for them and their families; it would benefit the entire country and the world.

So I am a great admirer of the California voters and their dedication to the future, their open minds, their willingness to think through complicated ideas even in the midst of a roaring pandemic — and to carry this over the top.

So many people worked so hard: patient advocate organizations, chambers of commerce, the university systems, the independent research institutes, cancer researchers, and supporters of science everywhere.

It was the civic society that led to this miraculous vote: that promises great things for the future in reducing human suffering: for you and I, for our families, and most importantly, for our children.

# Afterword: A Seeming Digression on Food Trucks

There is an energy which links us all and gives us energy beyond our numbers, when we unite in positive purpose.[1,2] (East Bay Times, https://www.eastbaytimes.com/2006/07/23/the-advocate/)

As you know, I loved working as a diver for Marine World Africa USA. It was glorious, being underwater with sharks, dolphins, eels, seals and killer whales.

Out back where the visitors seldom went, beside a winding salt-water canal, was a grass-thatched hut where we divers took our rest and patched the wetsuits.

And every few days, a shout would be heard, a rallying cry:

"FOOD TRUCK'S HERE! FOOD TRUCK!"

It had nothing to do with human lunch. But ours and the other buildings emptied, pouring out occupants; everyone stopped what they were doing and ran to help.

Before us was the Cut Shack, with stainless steel sinks and a huge freezer; where food was stored and prepared for our creatures.

The food truck held hundreds of boxes of restaurant quality food, to be unloaded and put away fast. Each box might have a thin film of ice to numb the fingers, but that would soon melt. Time was of the essence.

We formed a line, food truck to cut shack, and tossed the boxes to each other—pick up, pivot, pass along — to be stacked in sorted piles in the freezer room. Boxes of "Big Mac" (foot-long blue mackerel for the orcas) went in one corner; in the center was river smelt and squid, almost to the ceiling; steaks and livers for the lions piled up on the right.

Male and female, we were young and strong. We worked *fast*, lifting, tossing, laughing. In those brief moments, our strength and spirits joined; we became almost one. As the boxes flew along the line, we seemed no longer diver, trainer, welder, accountant; we were together — and felt capable of anything.

Quite suddenly, the job was done. We stood with empty hands a moment, looking at each other, as if puzzled by the loss of what had just been...

That spirit of cooperation is what it took to pass and renew the California stem cell program; and what it will take to continue such magic: not just in a single state or country, but wherever people suffer, and dream.

It takes one person to start, then many friends to get it done.

If you want to start a program in your state or country, I encourage you to do so; it is so well worth the doing. And maybe we can help each other, along the way.

There is of course only one Bob Klein, (and his son Rob Klein the third,) and I cannot volunteer their services.

But if you drop me a line, (address at the bottom of the page) I promise I will answer, assuming I am still alive.

Three more things. first, a vast resource. Right now, Americans for Cures Foundation is making a **giant archive** at Stanford University, with thousands of documents detailing how every challenge was met, to pass Prop 71 and Prop 14, to gain $8.5 billion in funding. The archive will be huge, the masterwork of Bob and Rob, Melissa, Mitra, Jacqueline, Anna, Elizabeth, U C Berkeley student Alison Lee and more; you will have no trouble Googling it.

Second, an Irish sheepherder **blessing** which President Ronald Reagan used to say, from the days when political opponents could still talk nicely to each other:

"May the road rise to meet you,
May the wind be always at your back;
May the sun shine warm upon your face,
And rains fall soft upon your fields.
And, until we meet again,
May God hold you in the palm of His hand."

And this above all: my personal motto — remember it when things go bad:

**If you never quit trying, you can only win, or die. And everybody dies — so why not try?**

My best to you, and to all whom you hold dear,

Don C. Reed

The author can be reached at: diverdonreed@gmail.com.

# Index